Amal Korrida

Nutrition, dietetics and diets

Amal Korrida

Nutrition, dietetics and diets

ScienciaScripts

Imprint

Any brand names and product names mentioned in this book are subject to trademark, brand or patent protection and are trademarks or registered trademarks of their respective holders. The use of brand names, product names, common names, trade names, product descriptions etc. even without a particular marking in this work is in no way to be construed to mean that such names may be regarded as unrestricted in respect of trademark and brand protection legislation and could thus be used by anyone.

Cover image: www.ingimage.com

This book is a translation from the original published under ISBN 978-613-8-40627-3.

Publisher:
Sciencia Scripts
is a trademark of
Dodo Books Indian Ocean Ltd. and OmniScriptum S.R.L publishing group

120 High Road, East Finchley, London, N2 9ED, United Kingdom
Str. Armeneasca 28/1, office 1, Chisinau MD-2012, Republic of Moldova, Europe
Printed at: see last page
ISBN: 978-620-5-98756-8

TABLE OF CONTENTS

General pedagogical objectives of the course: To be able to assess the nutritional needs of patients and vulnerable groups, to plan and adapt them, especially in the areas of dietetics, nutrition and therapeutic monitoring.

Section 1 : Nutrition and Dietetics

Specific educational objectives:

Theoretical objectives

S Define terms used in nutrition education

S List the different nutritional needs of humans

S Identify from a list the foods belonging to each food group

S Name the energy needs of the individual

S Establish the normal food ration from a qualitative and quantitative point of view according to age, sex and physiological state

S Describe the influence of diet on health

Practical and communication objectives

S Explain to a group of people the importance of nutrition supplementation

S Provide nutritional advice to different vulnerable groups: pregnant and breastfeeding women, infants and the elderly

Section 2 : Diets

Specific educational objectives:

Theoretical objectives

S Name the permitted and prohibited foods of the diets included in the course outline

S List the main signs of avitaminosis: A, B, C, D, K

Practical and communication objectives

S Explain the diet to clients on one of the diets listed in the diagram

S Explain to a client or a group of clients the symptoms of vitamin deficiencies: A, B, C, D, K, and how to prevent them.

The health of an individual is conditioned by a multitude of factors such as genetics, environment, stress, smoking, types of care, lifestyles, food consumption patterns, *etc.*

There is a strong correlation between nutrition and health, since a balanced diet is a real factor in the prevention of cardiovascular diseases, certain cancers, obesity and deficiency diseases.

In order for the health professional to participate appropriately in primary health care, in the promotion of correct nutrition, and to be able to integrate general and malnutrition-specific nutritional advice and recommendations (*i.e.*: over- and under-nutrition), it is important that he/she acquires knowledge related to the :

• Nutritional requirements

• Main characteristics of the food

• Risks of dietary deficiencies or excesses in vulnerable groups, such as women of childbearing age, the elderly, pregnant women and children

• Diseases with nutritional determinism, as well as means of prevention.

This course of nutrition is thus, programmed and elaborated to allow the future nurses a good control of these various nutritional concepts and also, to carry out targeted learning.

Nutrition: Discipline that studies food/feeding (*i.e.* all foods) and their use by the body.

Food: A substance that nourishes the body and ensures its growth and maintenance, and provides energy for its vital biological and physiological processes. With the exception of minerals and water, almost all foods are digested into simpler nutrients or nutrients.

Nutrient: A simple organic or inorganic substance that makes up food. Nutrients are digested and absorbed in the digestive tract and then used in the body's metabolic reactions. The most essential nutrients for the body to function are Proteins, Carbohydrates and Fats.

Dietetics: The study of hygiene and diet with emphasis on the nutritional value of foods.

Basal metabolic rate: Energy expended by a person at rest, in other words: it is the energy expenditure of a subject fasting, at rest (lying down), normally clothed, in a thermal environment of 20°C (temperature of thermal neutrality), and at emotional calm (no stress, no irritability).

The basic metabolism (BM) also corresponds to the energy needed to maintain life: heartbeat, breathing, nervous activity, maintaining a constant body temperature.

The contribution of the different organs to the MB varies according to their body mass and their metabolic activity (**anabolic** = set of reactions building large biological molecules from small ones by consuming energy in the form of ATP (Adenosine triphosphate) + **catabolic** = set of reactions degrading large biological molecules into small ones by providing energy in the form of ATP):

Organ	% contribution to body weight	% contribution to MB
Heart	1	9
Brain	2	19
Muscle	40	21
Liver	3	21
Kidneys	Less than 1	8
Digestive tract	2	1
Adipose tissue	21	3
Remainder	30	18

Nutritional supplementation: Consists of supplementing food with a specific food (vitamin, amino acid, mineral salt or antioxidant) in order to avoid and prevent a nutritional deficiency, for example following hair loss, fatigue, sunburn, stress, *etc.*

Calorie: Unit of measurement of the nutritional value of food (energy generated by its combustion in the body). It is also the amount of heat required to raise the temperature of one liter/kilogram of water by 1 degree Celsius (for example from 14.5°C to 15.5°C) at a normal atmospheric pressure of 101,325 Pascal.

By convention, the unit of measurement of body energy needs is the Kilocalorie.

1 Kcal = 1000 cal. Another alternative is the Joule (j). 1 cal " 4,18 d and therefore 1 Kcal " 4,18 Kj.

Metabolisable energy is proportional to the amount of food ingested, so when we give an intake for a gram of food, we note Kj/g and read kilojoule per gram.

Energy intake/value for 1 gram of :

- Carbohydrates = **17 Kj/g = 4 Kcal/g**

- Protein = **17 Kj/g = 4 Kcal/g**

- Fat = **38 Kj/g = 9 Kcal/g**

- Alcohol = **29 Kj/g = 7 Kcal/g**

Essential amino acids: 8 amino acids are essential for protein synthesis and the proper functioning of the body. The latter cannot manufacture them by itself, they are in this case : Lysine, Tryptophan, Phenylalanine, Leucine, Isoleucine, Valine, Methionine, Threonine.

A good proportion between these amino acids is necessary for their good assimilation. Indeed, if an amino acid is in very low quantity in our food ration, it is the whole which will be little absorbed. This amino acid is called the **limiting factor**.

Food ration: Daily quantity of food necessary to an individual to cover its energy needs (building blocks such as proteins) and plastic needs (such as calcium ensuring the maintenance and manufacture of living matter, *e.g.*: replacement of dead cells, repair of damaged tissue). The food ration must be sufficient to achieve a certain balance between the different components (Carbohydrates, Proteins and Lipids) and varied according to the age, physical activity and physiology (growth, illness, pregnancy, breastfeeding) of each individual.

Nutritional requirements: A set of nutrients that provide the body with the energy and elements necessary for maintenance, physiological and metabolic functioning, as well as the development of each cell in a healthy individual (homeostasis).

These needs vary according to age (growth, aging), sex (♂, ♀), physical activity, physiological conditions (gestation, lactation), *etc*.

Avitaminosis: A condition caused by a vitamin deficiency in the diet (e.g. Biermer's anemia, due to an anomaly in the assimilation/absorption of vitamin B12 in the stomach, beriberi, due to a vitamin B1 deficiency, rickets, due to a lack of vitamin D in the body).

Antioxidant : Molecules that naturally capture free radicals and protect cells against aging. Antioxidants play a role in hypocholesterolemia and the prevention of cancers and cardiovascular pathologies. The family of antioxidants includes β-carotene (lycopene and provitamins A), ascorbic acid (vitamin C), tocopherol (vitamin E), polyphenols constituting flavonoids (substances widely found in plants), tannins (in cocoa, coffee, tea, grapes, *etc.*), anthocyanins (especially in red fruits) and phenolic acids (in cereals, fruits and vegetables).

1.   Health consequences of a good diet

Balanced food gives our body the information, energy and elements necessary for its proper biological and physiological functioning. It fulfills digestive, respiratory, circulatory, excretory and endocrine functions allowing the supply to the cells of elements necessary to their growth, the progress of the various metabolisms, as well as the elimination of their waste.

In the absence of this, the body's metabolic and energetic processes suffer and our health deteriorates.

2.   Health consequences of inadequate nutrition

If the daily food intake is disproportionate, 2 cases can occur: either under-nutrition (food deficiency, starvation), or over-nutrition (obesity).

a.   *Dietary deficiencies*

If nutritional needs are not met, nutritional diseases can occur. These include:

• Protein-energy malnutrition in children, leading to marasmus and kwashiorkor.

• Micronutrient or trace element deficiencies, namely:

Vitamin A deficiency, causing night blindness or night blindness.

Vitamin C deficiency, causing scurvy.

Vitamin B1 deficiency, causing beriberi.

Vitamin B5 or PP deficiency, causing pellagra.

Vitamin B9 deficiency, causing neural tube defects.

Vitamin D deficiency, causing rickets.

Vitamin K deficiency, causing blood clotting disorders.

Vitamin E deficiency, causing reproductive disorders.

Iodine deficiency, causing goiter, dwarfism, and cretinism.

Iron and folic acid (vitamin B9) deficiency, causing anemia.

b.   *Excess food*

Excessive eating can also lead to nutritional disorders and cause a number of pathologies such as hypercholesterolemia, diabetes mellitus, obesity, cardiovascular diseases, arthritis, gout, *etc.*

Therefore, what we eat is crucial to our health and well-being. It is said that food acts as medicine, to maintain health, prevent, and treat diseases. For example, food contributes to the development, management and prevention of mental health problems such as: depression, schizophrenia, attention deficit hyperactivity disorder, and Alzheimer's disease.

IV. Food group classifications

In nutrition, different methods of classifying foods have been proposed. Foods can be grouped according to their :

Constituents in major nutrients: in macronutrients: Carbohydrates, Lipids, Proteins (GLP), and micronutrients: vitamins, minerals.

Nutritional values: for example, energy foods = carbohydrates and fats, foods with a functional/protective role such as mineral salts + vitamins, and building foods that serve for the constitution of the body such as proteins in addition to calcium.

Chemical compositions: carbohydrates (sugars or bones or carbohydrates), proteins and amino acids, lipids and fatty acids, vitamins and minerals. Dietary fiber and water are sometimes added to this list.

Commercial values: for example, cereals, poultry, roots and tubers, nuts and oilseeds, meats, fish, fruits and leafy vegetables.

In this course, we will classify foods according to the 7 food groups that make up the food pyramid and according to the major nutrient constituents.

J- Classification according to the food pyramid

To better balance the food intake of our meals, it is necessary to know the different food groups. The classification of foods into groups is based on the nutritional value of the nutrients in the food. This will allow a better identification of the food profiles potentially inducing deficiencies, and will facilitate the presentation of guidelines to patients in terms of administration of the different foods. Seven food groups will be studied in this course.

Group 1: Sweetened products (the top of the pyramid)

- Chocolate

- Honey

- Pastries

- Confectionery

- Dried fruits: grapes, dates, figs, prunes

- Sugar

Generally considered as pleasure foods, they provide simple sugars that are quickly absorbed and assimilated by the body.

Examples of sugar equivalents:

1 sugar = 5 g

1 can of soda = up to 8 sugars!

The consumption is to be limited to less than 10% of the food ration.

Recommended Daily Allowance (RDA): 1 portion of 10 g/meal maximum.

Group 2: Fat

- Butter

- Cream

- Oils: olive, argan, peanut

- Smen

- Margarine

- Almonds

They bring lipids and liposoluble vitamins (A and E). They can be of animal or vegetable origin.

Saturated fatty acids will be the main component of animal fats, while those from the plant world will provide mostly unsaturated fatty acids.

Among the fats of vegetable origin that are sources of vitamin E, there are monounsaturated (Omega 9) and polyunsaturated (Omega 3 and 6) fatty acids:

- oils of: corn, olive, almond...*etc.*

- vegetable margarines

Among the animal fats that are sources of vitamin A and D:

- butter

- fresh cream

For lipids of animal origin, there are saturated fatty acids which represent a source of cholesterol

Consumption should be limited for saturated fatty acids of animal origin and cholesterol. **AJC:** 1 to 2 portions of 10 g of animal fat maximum.

3 to 4 servings of vegetable fat maximum.

Group 3: Meat, fish, eggs

- Meat: beef, lamb, goat, camel, poultry (rooster, turkey, guinea fowl, *etc.*)

- Meat offal (tongue, heart, liver)

- Poultry offal (heart, gizzard, liver)

eggs: chicken, duck, goose, quail, pheasant, pigeon

- Delicatessen: Kosher, ham, sausage, Kaddid, Khliaa,

- Game : birds (partridge, bustard, pigeon, turtle-dove), gazelles, hare, doe

- Fish: carp, sardine, pigeon, sole, skate, monkfish, anchovy, whiting

- Shellfish: mussels, scallops, oysters

- Crustaceans: Shrimp, lobster, crab

- Cephalopods : Squid, sepia, cuttlefish, octopus

They provide good quality animal proteins that are better assimilated by the body than plant proteins, because they are composed of all amino acids, including the 8 essential amino acids, in addition to iron and polyunsaturated fatty acids ($\Omega3$ in fish and seafood), in fact :

- Red meat provides haem iron which is better absorbed by the body because it is bound to haemoglobin.

- Fish proteins are of high biological value and nutritional interest. Fish are also a source of certain essential fatty acids, especially fatty fish rich in $\Omega3$ and $\Omega6$ like sardines, trout, tuna, anchovies, mackerel, salmon. *etc.*

- Eggs contain albumin, which is also considered a highly digestible and easily assimilated protein of high biological value. Egg yolks, however, contain a lot of lipids (fats) which can make the whole egg less digestible.

ô Consumption should be limited for saturated fatty acids.

AJC: 1 to 2 servings of 100g/d, depending on age, physiological condition and physical activity.

Fish: **2 to 3 times/week**, including 1 fatty fish (*e.g.* sardines)

Group 4: Dairy products

- Milk: cow's, camel's, goat's, breastfeeding, UHT, pasteurized, raw, sterilized, skimmed, powdered, whole

- Milk derivatives: Lben, yogurt, milkshake, ice cream, rayeb

- Cheeses : Mozzarella, Jben, Roquefort, square

They provide good quality animal proteins, vitamins (A, D, B), and calcium linked to phosphorus and casein, a milk protein.

Consumption should be limited for salty dairy products such as cheeses.

AJC: 3 servings/d of 30g of cheese or 1 yoghurt or 200 ml of milk.

Group 5: Fruits and vegetables

- Raw fruits: Orange, kiwi, apple, banana, strawberry

- Vegetables: raw (turnip, broccoli, zucchini, asparagus, onion, cauliflower, artichoke, lettuce, spinach, bell pepper...*etc.*) and dried (bean, pea, lentil...*etc.*)

- Frozen, raw (raw vegetables), cooked (cooked vegetables) or processed (soups)

They provide vitamins (for example: vitamin C for raw fruits and vegetables), minerals, fiber, water and a quantity of carbohydrates.

- Vegetables provide only 25 Kcal (kilocalories) on average per 100 grams. They make it possible to add volume to the meal without calories, as long as they are not cooked with too much fat. They therefore promote satiety (= satisfaction and saturation *vs.* hunger) while promoting intestinal transit through the fiber and water they contain.

- Fruits provide an average of 50 Kcal per 100 g (100 Kcal for the banana). They complete the vegetable intake by bringing a "touch of pleasure" with their sweet taste.

The consumption is to be limited for fruits very rich in simple sugars.

CDA: at least 5 servings of 80 g/day, i.e. 400 g, with 2 servings for raw fruits and vegetables to provide the body with the necessary vitamin C.

Group 6: Starchy foods

- Pulses (dried vegetables): Lentils, peas, dried beans

- Cereals or seeds: Corn, wheat, flax, rice, barley

- Tubers (underground organs): Potato, sweet potato

- Finished industrial products: Bread, flour, spaghetti, chips, pasta, french fries

They provide carbohydrates of plant origin with slow assimilation (starch, complex sugars).

Cooked starchy foods provide an average of 100 Kcal per 100g. Bread, rusks and cereals contain less water, so they are more concentrated in carbohydrates and have more calories

AJC: at each meal and according to age and physical effort. In general, whole and semi-complete cereals (wholemeal bread) and legumes are preferred.

<u>Group 7: Beverages (the base of the pyramid)</u>

- H_2 O: Carbonated, still (not sparkling), mineral, tap

- Infusion: Mint tea, herbal tea (star anise, verbena, lemon balm ... *etc.*)

- Peach Iced Tea

- Coffee: Arabica, Cappuccino, Starbucks

- Juice/Cocktail

- Lemonade/Lemonade

- Soda: Coca Cola, Red Bull

- Alcohol/vinegar

They bring water, minerals, fast sugars, alcohol.

The human body is composed of 2/3 of water, that is to say ~ 65% of the total body mass, the contributions in liquid are thus vital and essential.

AJC: At least 2 large glasses of water with each meal, or about 1.5 l/d

This depends on the physical activity, the state of health, the climatic conditions, the age, *etc.*

<u>**N.B.:**</u> After their degradation and digestion, foods are transformed into nutrients, thus :

Carbohydrates and sugars will give glucose molecules

The proteins will give amino acids

The lipids will give fatty acids

The dietary fiber will not be digested, but will facilitate the food transit.

7 Food Groups	FoodFood	Composition of Nutrients	Roles of Nutritional Intakes

61: Carbohydrates and sugars		Sugar, honey, chocolate, jam, syrup, pastries	Carbohydrates (sugars that are quickly absorbed by the body)	Source (Teneme par excellence
G2: Fat		Butter, cream, oil, margarine	Lipids Vitamin A for butter Vitamin E and essential ms acids in oils and marsarines	Very essential (energy and reserve elements)
G3: Meat, fish, eggs		Crustaceans, poultry, beef meats, sheep, rabbit, goat	Quality proteins, iron, vitamins B and D, low fat (except for cold cuts)	Maintenance of muscles, skin, blood (Nutrient builders)
64: Dairy products		Milk, cheese, yoghurt, dairy desserts, ice cream	Protein, vitamins especially A and D, calcium, fat	Construction and development of bones and teeth (Building nutrients)
65: Fruits and vegetables		All fruits, all green vegetables	Fibers, sugars, vitamin C	Vitality and fight against diseases and constipation (facilitate intestinal transit)
66: Starchy foods, cereals, pulses		Bread, rice, pasta, potatoes, lentils, dried beans, chickpeas, broad beans	Carbohydrates (slow sugars), proteins, B vitamins, minerals, fiber	Energy required for physical activities and intense efforts
67: Drinks		Water (mineral, sparkling, tap), tea, infusions, coffee, herbal teas, juice	Water, sugar, mineral salts	Vital and moisturizing

4- Classification according to major nutrient constituents

A. Macronutrients (GLP):

1. Proteins and amino acids (Building blocks)

1. **Proteins are** made of Carbon, Hydrogen, Oxygen and Nitrogen (CHON). They represent 15% of the human body mass, are constantly renewed and are found in the skin, muscles, hair, bones and almost all metabolisms (hormones, enzymes... *etc.*).

Roles of proteins in biological processes:

- Enzyme catalysis: since enzymes are protein molecules that catalyze almost all chemical reactions in biological systems (chromosome replication, hydration of CO_2).

- The transport and storage of ions and biological molecules: for example, iron which is transported in the plasma by the transferrin protein and stored by the ferritin protein in the liver, or the transport of O_2 in the erythrocytes (red blood cells) by the hemoglobin protein, or the transport of O_2 in the muscles by the myoglobin protein

- The coordination of movements, in the case of muscle contraction, which is made possible by actin and myosin filament proteins, or the movements of chromosomes during mitosis, or the propulsion of spermatozoa made possible by contractile protein assemblies in the flagella.

- Mechanical support: mechanical resistance to stretching and traction by collagen and fibrous proteins (elastin and keratin).

- Immune protection: antibodies.

- The generation and transmission of nerve impulses at the synapses (junction between 2 nerve cells): the response of nerve cells to different *stimuli* is done *via* receptor proteins such as: acetylcholine (neurotransmitter).

- Control of cell growth and differentiation: the expression of genetic information must be controlled to ensure the development and differentiation of the different cells of the organism.

2. <u>Amino acids</u>

Proteins are made up of several molecules or basic structural units called amino acids (aa). In humans, 20 aa are involved in the different metabolic pathways, of which 8 aa (9 in newborns) cannot be synthesized by the body, and therefore the latter must provide them through food. These are **essential amino acids:** Leucine, Isoleucine, Valine, Lysine, Methionine, Phenylalanine, Tryptophan, Threonine, in addition to Histidine in infants.

The nutritional value of a protein is proportional to the different essential amino acids it contains. The more essential amino acids a protein contains, the higher its nutritional value and the better its quality. If an amino acid is in very low quantity in the food ration (i.e. impossible to be used in the construction, the repair or the cellular and tissue renewal), it is the whole of the aa which will be little assimilated. This amino acid is called the **limiting factor**. Example: tryptophan is often scarcer in our modern diet, so it can prevent the assimilation/absorption of other amino acids and promote deficiencies.

Each food does not have a perfect distribution of essential amino acids, **except for** the egg white, which is considered the reference protein par excellence.

Origins and sources of proteins and aa :

Proteins can come from animals (fish, poultry, cows, beef, in the form of meat, dairy products and eggs), or from plants (vegetables and fruits, legumes: chickpeas, soybeans, lentils, dry beans, etc., cereals: wheat, barley, oats, etc., and oleaginous fruits: sunflower seeds, almonds, nuts, hazelnuts, etc.). *etc.,* cereals: wheat, barley, oats, *etc.,* and oleaginous fruits: sunflower seeds, almonds, walnuts, hazelnuts, *etc.*).

Animal proteins generally have all the essential amino acids needed to make other proteins after digestion, which is not the case for those of plant origin. Certainly, cereals are generally poor in Lysine, while oleaginous fruits are deficient in Lysine, and legumes contain little Methionine and

Cysteine.

In order to remedy this lack, it is advisable to <u>supplement our food with proteins</u>. There are two choices:

a.   Adding animal and vegetable proteins in the same meal (cheese and bread)

b.   Add other vegetable proteins that can provide missing amino acids, especially when the food and agricultural resources of a country are very limited (poverty). For example, in India, they add rice to dhal (lentil puree) in the same meal, in Italy: beans to pasta, in North Africa: durum wheat semolina to chickpeas.

Daily Nutrient Requirements for Protein or RDA:

The recommended dietary allowance (RDA) for protein is between **12 and 15% of total Kcal** (from 2500 Kcal), or on average for a moderately active adult, 70 g of protein/day or **0.8 g/Kg/d.**

	Infants (1-12 months)	Infants (4-15 years old)	Teenagers (15-19 years old)	Adults (19-65 years)
Woman	1.1-2 g	0.9-1 g	0.Sg	O.Bg
Male			0.9 g	

For a balanced diet, it is therefore necessary to establish protein equivalents (70 g/d for a normal adult male and 52 g/d for a normal adult female), according to the foods making up the 3 daily meals, as shown in the following example, taken from the Swiss Society of Nutrition (2004):

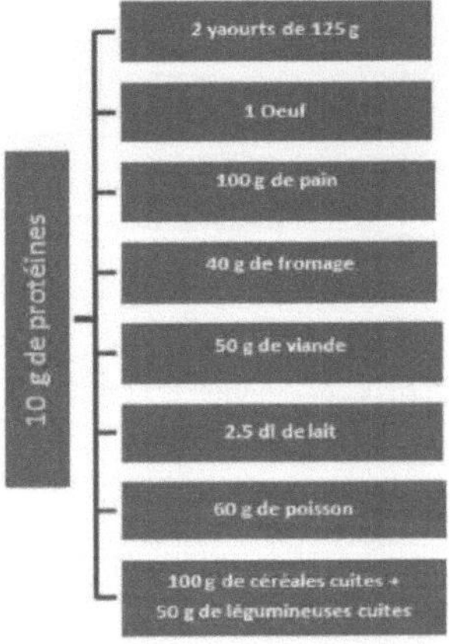

<u>**N.B.:**</u>

In practice, it is often difficult to respect the RDA (12-15% of total Kcal) or the RDA (0.8g/Kg/d), which are theoretical values, especially in at-risk populations: pregnant and breastfeeding women, strict vegans with a diet frequently low in essential aa and vitamin B12, vegetarians, whose diet is deficient in animal proteins, anorexic people (models, teenagers, old people, *etc.*).

The figure below illustrates the nutritional needs of normal adults, for the needs of vulnerable people such as pregnant and lactating women, the elderly, *etc.*), they will be presented in section VI of this course.

Summary:

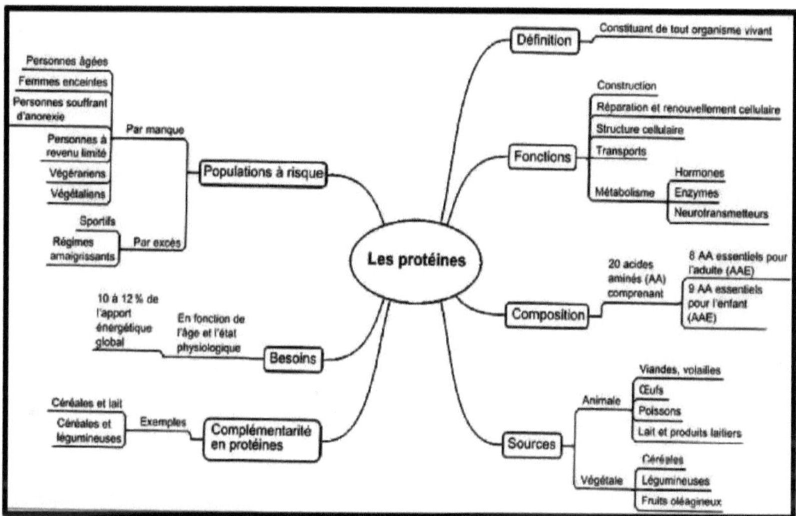

II. Lipids: Cholesterol, Triglycerides and Fatty Acids: (Energy Nutrients)

Lipids are made up of Carbon, Hydrogen and Oxygen (CHO). They represent 17% of human fat mass and are composed of the family of fatty acids, the main constituents of triglycerides, phospholipids, waxes, steroids (cholesterol), fat-soluble vitamins and terpenes (menthol).

Roles and characteristics of lipids in the body:

- These are the most energetic food compounds (9 Kcal/g), whose RDAs represent 30 to 35% of the daily TEA

- They constitute the cell membranes and nervous tissue (phospholipids, cholesterol)

- They are considered an important energy reserve (triglycerides) that protect the main organs of the body and are stored in the adipose tissue

- They are precursors of molecules regulating cellular functions (synthesis of steroid hormones)

- They facilitate the absorption of fat-soluble vitamins such as vitamin E, D, K and A

- They are insoluble in water and aqueous media (= generally hydrophobic)

- Being hydrophobic, lipids are transported and carried in the bloodstream by water-soluble lipoproteins (lipid and protein molecules), such as chylomicrons, which are formed during digestion and are responsible for the transport of lipids from the small intestine to the peripheral adipose tissue where they are reprocessed.

The lipid family

It is composed of the :

1. Cholesterol: a lipid that is carried in the bloodstream by 3 types of lipoproteins: Low density lipoproteins (LDL or "Low Density Lipoprotein" = like Buses that carry bad cholesterol to the arteries, forming atherosclerotic plaques), very low density lipoproteins (VLDL or "Very Low Density Lipoprotein"), and high density lipoproteins (HDL or "High Density Lipoprotein" = like Taxi carrying good cholesterol without clogging the arteries).

2. Triglycerides (TG): are formed by 3 molecules of fatty acid (GA) + 1 molecule of glycerol. They constitute 95% of dietary lipids.

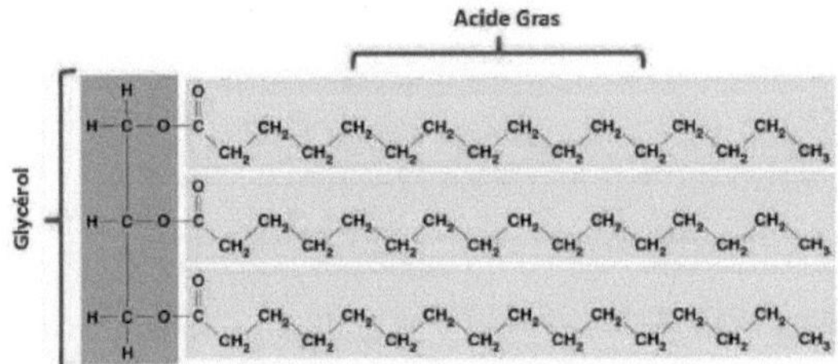

1 Glycérol + 3 Acides Gras = 1 Triglycéride

Food absorption in the mucus and intestinal villi is facilitated by the pancreatic enzyme <u>lipase</u>, which degrades TG into 1 glycerol and 3 GAs.

3. Fatty acids (FA): are linear chemical compounds of the type :

Saturated fatty acids (SFAs): aliphatic carboxylic acids consisting of 12 to 24 carbon atoms and <u>no carbon-carbon double bonds</u>: all carbon atoms are saturated with hydrogen, e.g., pamlitic acid: $CH_3(-CH_2)14-COOH$. Generally, saturated fatty acids are hypercholesterolemic (produce a lot of LDL), and are a major cardiovascular risk factor.

Unsaturated fatty acids (UFA): produce a lot of HDL and represent the most important part of consumed fatty acids. They have <u>1 carbon-carbon double bond </u>for <u>monounsaturated fatty acids</u>

(MUFA), such as oleic acid (from olive oil): CH₃ (-CH₂)7- CH=CH(-CH₂)7-COOH, or <u>several carbon-carbon double bonds</u> for <u>polyunsaturated fatty acids (PUFAs),</u> such as linoleic acid of the Omega 6 family: CH₃ (-CH)23 (-CH₂ - CH=CH)2(-CH2)7-COOH or alpha-linolenic acid from the Omega 3 family: СЩ-CH2- CH=CH)3 (-CH)27 -COOH. The last 2 GA are called <u>essential fatty acids (EFA),</u> because they can not be biosynthesized by the body, and therefore, must be fully provided by food.

N. B: The mode of action of these fatty acids is complex. Omega 6s play roles in reproduction, epidermal integrity, blood coagulation, regulation of lipemia (cholesterol), as well as the immune and inflammatory systems and their metabolites (can therefore be pro-inflammatory, pro-thrombotic and hypertentive).

For the metabolites resulting from the GA of the Omega 3 family, it is the opposite (they are anti-inflammatory, anti-thrombotic, and they often lower the blood pressure). An excess of omega 6 compared to omega 3 would tend to favor the development of various diseases such as: cardiovascular diseases, cancers, and various inflammatory and autoimmune diseases.

Figure: Classification of GAs (Examples and sources)

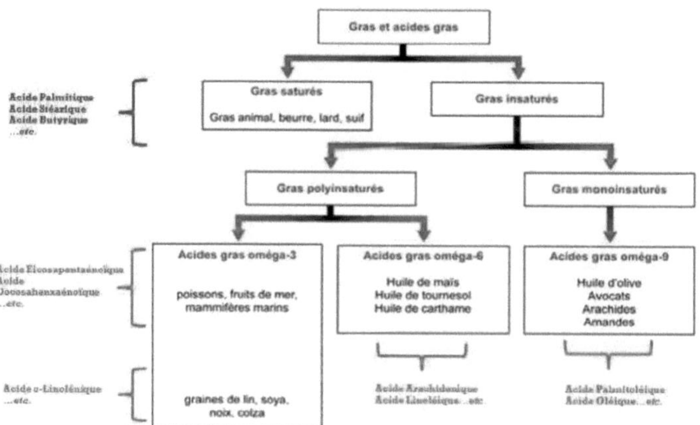

Figure: PUFA family: Omega 3 and 6

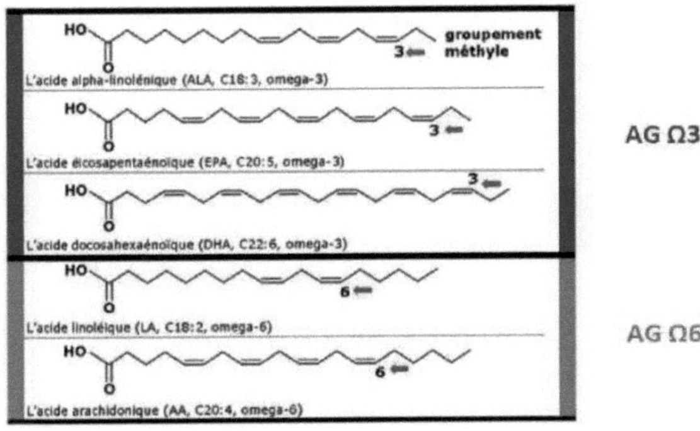

***Trans* and *cis* fatty acids (will not be treated in detail in this course):** generally, they are GA with different spatial configurations (allosteric), and like saturated GA, they cause cardiovascular pathologies.

Sources and origins (animal + vegetable) of lipids :

Acide Gras	Sources	Effets sur la santé
AG saturés	Viandes, charcuterie, laitage, graisse de palme et de coco	Source d'énergie Effet hypercholestérolémiant
AG mono-insaturés (AG Ω 9)	Huiles: d'argan, d'olive, de colza, d'arachide, amandes, avocat, noix	Source d'énergie Effet protecteur des vaisseaux + du cœur
AG poly-insaturés		
Ω 3	Huile: de soja, de noix, de colza, graines de lin, poissons gras, fruits de mer	Précurseurs de synthèse hormonale tissulaire (immunité) Remplacent les AG saturés et sont favorables pour la circulation sanguine Indispensables à la croissance du cerveau (apprentissage, mémoire, concentration) Boostent les défenses immunitaires + ont des propriétés anti-inflammatoires Contribuent au bon fonctionnement cardio-vasculaire Favorables au bien-être mental (antidépresseur)
Ω 6	Huile: de tournesol, de hareng, de pépins de raisins, de carthame, de maïs, la graisse d'oie, les germes de blé	Quand équilibrés, ils traitent l'hypercholestérolémie En excès, ils conduisent à: * l'obésité * les maladies cardio-vasculaires * l'asthme (immunité) * la thrombose (agrégation plaquetaire + formation du caillot sanguin ou thrombus) * l'arthrite (inflamation articulaire)

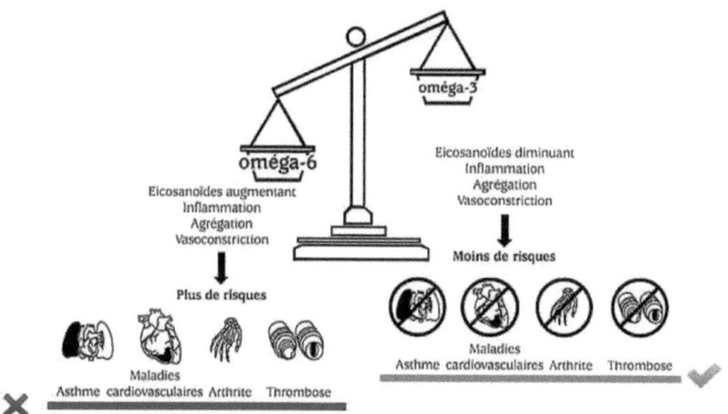

Recommended Daily Allowance for Fats or RDA :

The fats in our diet represent 80% of <u>hidden fats</u> such as: meat, fish, and cheese, and 20% of <u>non-hidden or visible fats</u> that we add to our meals such as: oils, butter, *etc.*

The energy released by the complete oxidation of GA is 9 Kcal/g.

A balanced diet includes an amount of fat between **30 and 35% of the total** daily <u>**energy intake (TIE)**</u>, which is on average 2500 Kcal/d.

The recommended distribution for :

Saturated fatty acids, is 8% of the TEA,

Monounsaturated fatty acids, is 20% of the TEA,

Polyunsaturated fatty acids, is 5% of the TEA,

$$\text{Avec le rapport :} \quad \frac{Acide\ linoléique\ (\Omega\ 6)}{Acide\ \alpha-linolénique\ (\Omega\ 3)} \geq 5$$

In case of sedentary life (no or limited physical activity), carbohydrates usually provide the energy necessary for the functioning of the body. However, during long periods of energy expenditure, the body is obliged to look for energy in lipids, if the carbohydrate intake is insufficient.

<u>Remark:</u>

It is important to choose the fats with which you cook, for example, it is advisable not to heat too much olive oil, rapeseed oil or soybean oil, because they break down quickly, unlike sunflower oil known as "High oleic", peanut oil or refined olive oil which are more resistant to high temperatures between 180 and 190°C.

Summary:

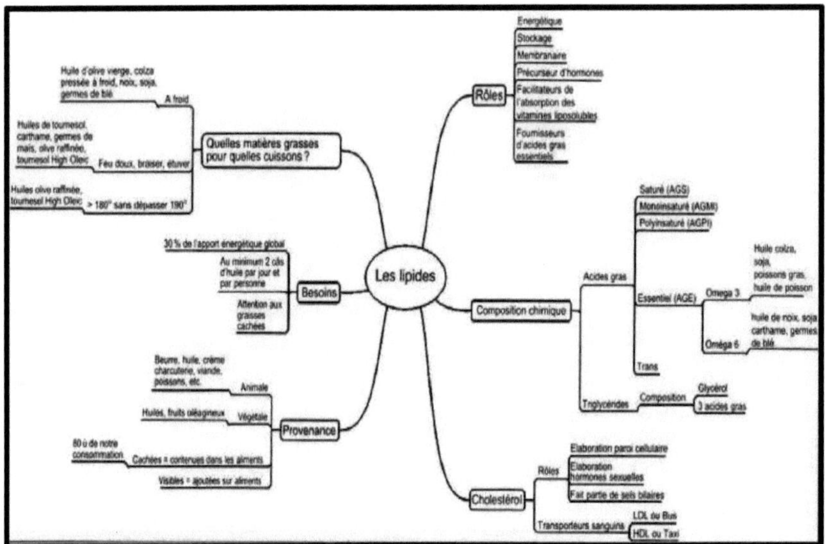

111. Carbohydrates (Energy Nutrients)

Carbohydrates, sugars, oses or carbohydrates are made up of Carbon, Hydrogen and Oxygen (CHO). These energy molecules, like proteins and lipids, can have an animal or vegetable origin, and they are used by almost all the cells of the body in the form of glucose. However, their storage is very limited and the body only stores a small amount in the form of glycogen in the liver and muscles. If carbohydrates are needed, the body can synthesize them from proteins *via* gluconeogenesis or from lipids via ketogenesis (these two syntheses will not be covered in this course).

This is why there are no essential carbohydrates (as for amino acids or fatty acids).

The carbohydrate family includes the :

Monosaccharides or simple sugars with 1 glucose molecule such as: fructose (fruit and honey sugar), galactose (milk and yogurt sugar), and glucose.

Disaccharides or simple sugars with 2 glucose molecules such as: sucrose (table sugar, composed of 1 glucose + 1 fructose molecule), lactose (= 1 glucose + 1 galactose), and maltose (= 2 glucose molecules).

Polysaccharides or complex sugars with *n* glucose molecules such as: starch, glycogen, cellulose, pectin, lignin (these last 3 are of vegetable origin and are not digested by the digestive juice, nor by the bacteria of the colon, because they only facilitate the food transit. They are dietary fibers).

<u>Remark</u>:

Sucrose bound to cooked starch, galactose, and lactose are fermentable carbohydrates and can become cariogenic (responsible for the formation of dental caries).

Roles and characteristics of carbohydrates in biological processes:

- Glucose plays the role of fuel that provides energy in the form of ATP (1 molecule of glucose provides 38 molecules of ATP and 1g of carbohydrates = 4 Kcal).

- Carbohydrates are stored as glycogen in the liver and muscles

- Pentose sugars (with 5 carbon atoms) are used in the composition of nucleic acids (RNA), glycoproteins and glycolipids

- Carbohydrates are essential for carbohydrate-dependent cells (red blood cells + brain + renal medulla)

- Carbohydrates are characterized by their glycemic index (GI), which is a criterion for classifying carbohydrate-containing foods: based on their effect on blood glucose levels during the 2 hours following ingestion. In the case of pathologies such as obesity and diabetes, the GI makes it possible to compare the glycemic power of each food, which is measured directly during digestion. The glycemic index of a food is given in relation to a reference food, which is given an index of 100 (usually pure glucose or white bread).

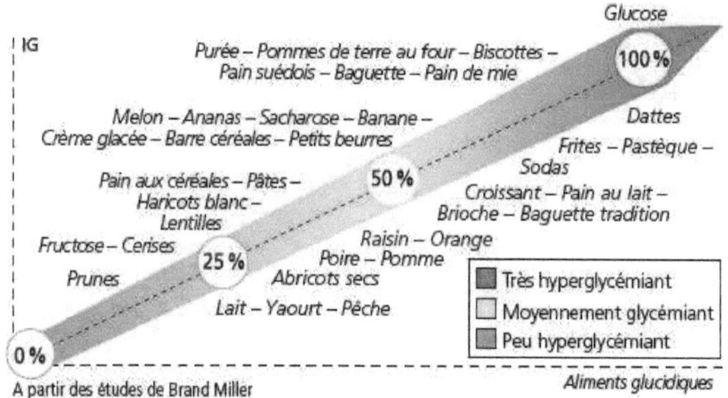

Daily energy and nutrient intake for carbohydrates (RDA):

1 g of carbohydrates provides 4 Kcal of energy.

A balanced diet includes between 50 and 55% of the energy ration in carbohydrates.

Nutrition course (ISPITS Agadir - Annex of Tiznit)

Summary:

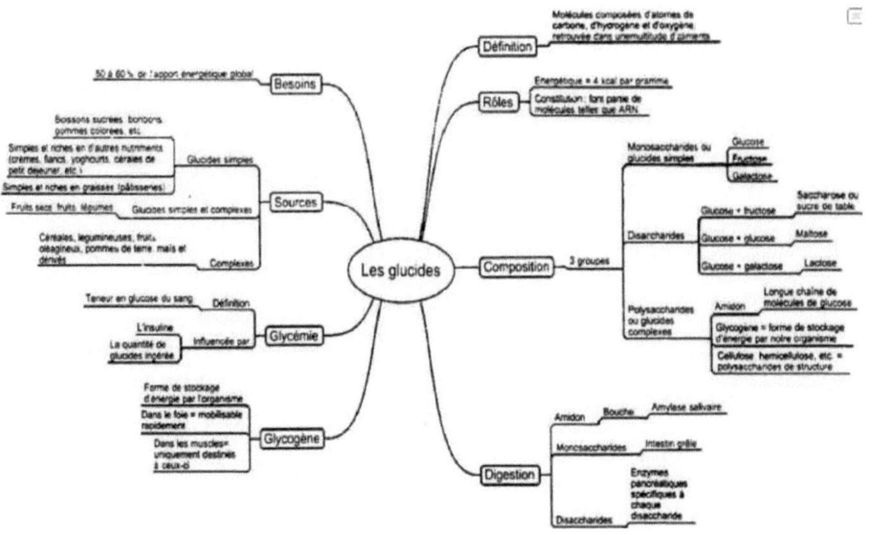

	Food source	Chemical components	Role & Function in the organization	Nutritional value	Energy requirements RDA (for adults)	Pathological implications
Protein	**Animals:** Fish, Poultry, Meat, Dairy products. **Plants:** Pulses Cereals	~ 20 Amino acids (8 of which are essential) + Peptides	Growth + regeneration (Builders)	**Quantitative:** **Qualitative:** By the 8 essential aa: composition of hormones, enzymes, antibodies	*1g = 4 Kcal *12-15% of daily caloric intake *0.8g/Kg/d	Malnutrition Hyperuricemia Renal insufficiency
Carbohydrates	**Animals:** Honey, Dairy products. **Plants:** Beet Cereals Sugar cane	Mono, di, and polysaccharides, Glucose, starch, fructose, galactose...	* Provide energy mainly in the form of ATP * Constitution of glycogen reserves in the muscles + liver	**Quantitative:** Energy intake in 1er location **Qualitative:** Composition of red blood cell membranes	*1g = 4 Kcal *50-55% of daily caloric intake	Obesity Diabetes Caries

			* Synthesis of DNA, RNA and coenzymes (Energetic)			
Lipids	**Animals:** Fish, Poultry, Meat, **Plants:** Oils Cereals	Fatty acids Cholesterol, Phospholipids, Triglycerides,	* Synthesis of steroid hormones * Constitution of reserves in adipose tissue (Energetic)	**Quantitative:** **Qualitative:** Composition of cytoplasmic membranes, brain, nervous tissue	*1g = 9 Kcal *30-35% of daily caloric intake	Obesity Hypercholesterolemia Dyslipidemia Infarction Diabetes type II Intestinal malabsorption

1. Micronutrients (Minerals and Vitamins)

These are nutrients that are essential to the proper functioning of the body, and have no energy value. They exist in very small quantities or traces (ug or mg), and include about ten minerals and 13 vitamins.

Roles and characteristics of micronutrients in biological processes:

They are at the origin of certain enzymes, coenzymes and hormones

They are involved in the acid-base balance

They play the role of catalysts for different metabolic reactions

They are involved in the water balance (osmotic pressure)

They are part of the structure of bones and teeth

... *etc.* (see tables of macro and µelements)

1. Mineral salts

The important mineral elements or macroelements are :

Sodium (Na), potassium (K), chlorine (Cl), calcium (Ca), phosphorus (P), and magnesium (Mg).

The trace elements are: iron (Fe), zinc (Zn), copper (Cu), manganese (Mn), iodine (I), selenium (Se), chromium (Cr), fluorine (F), molybdenum (Mo), cobalt (Co), silicon (Si), vanadium (V), nickel (Ni), boron (B), and arsenic (As)

The following table groups together the majority of mineral salts: their origin, ANC, functions and diseases in case of lack in the body:

Minerals	Food/Feed	ANC/d	Function and role	Pathologies in case of deficiency

MACROELEMENTS

Sodium (Na)	Table salt, fresh fish, canned fish, gherkin, khliaa,	5000 mg	Acid-base balance of the body (pH) Hydration In case of excess: increases blood pressure, stroke and osteoporosis	Drop in blood pressure Confusion Loss of appetite Muscle cramps
Potassium (K)	Banana, yogurt, spinach, squash, potato, clam, tomato, dry bean	4700 mg	Maintenance of the acid-base balance of the body Muscle contraction Transmission of nerve impulses Function of the kidneys and adrenal glands	Muscle cramps Fatigue Constipation Bloating Hypokalemia
Magnesium (Mg)	Chocolate, nuts, seeds, artichoke, cereals, vegetables, legumes, almonds	300-400 mg	Nerve transmission (synapse) Muscle relaxation after contraction 300 metabolic reactions	Loss of appetite Muscle cramps Fatigue and migraine Irregularity of the heart rhythm Nausea and vomiting Coronary spasms
Phosphorus (P)	Fish, seeds, poultry, dairy products, soft drinks, dried vegetables	700 mg	Composition of DNA and RNA Composition of teeth and bones Acid-base balance of the body	Loss of appetite Muscle cramps Fatigue and migraine Bone demineralization Tingling and pins and needles Confusion Bone pain
Chlorine (Cl)	Table salt	1-2.5 g	Maintenance of osmotic balance Gastric secretion (HCl) Constituent of the cerebrospinal fluid	Diarrhea or vomiting Muscle cramps Hypochloremia
Calcium (Ca)	Fish, dairy products, dried vegetables, nuts, oil seeds, fruits and vegetables	1 g	Composition of teeth and bones (Growth)	Osteoporosis Renal pathologies Dentition and gum problems Premenstrual

				syndrome

Minerals	Food/Feed	ANC/d	Function and role	Pathologies in case of deficiency
OLIGOELEMENTS				
Zinc (Zn)	Liver, oyster, cheese, lobster, coriander, cumin	10-12 mg	Immunity Synthesis of nucleic acids and proteins	Immune problems Anorexia
	meat, bread		Modulation of mood and learning	Male fertility problem)
			Neurological functions	Fetal malformations
			Reproductive functions	Skin lesions
			Enzymatic reactions	
Iodine (I)	Salt, fish, dairy products, beans	150 µg	Thyroid hormones	Goiter
Iron (Fe)	Spinach, liver, cereals, meats, shrimps	9-16 mg	Hemoglobin synthesis in red blood cells Regulation of cell differentiation Production of ATP	Oxidative stress Hemochromatosis
Copper (Cu)	Artichoke, oysters, mushrooms, shrimps, lentils, beans, molds, dark chocolate	1.5-2 mg	Enzyme synthesis Antioxidants (free radicals) Regulation of neurotransmitters Bone and cartilage metabolism Iron metabolism (Hemoglobin synthesis)	Hepatitis Hemolytic jaundice
Fluorine (F)	Cabbage, spinach, fish, (Mother's milk is low in it!) *H2O* mineral	2-2.5 mg	Composition of the teeth	Dental and bone fluorosis (caries) Vomiting and headaches Abdominal pain
Chromium (Cr)	Broccoli, liver, dry bean, plum, mushroom, yeast, meat, asparagus	55-65 µg	Insulin co-factor	Disorders of metabolism of GLP
Molybdenum (Mo)	Liver, dairy, dark green vegetables, cereals, dried vegetables	30-50 µg	Amino acid metabolism Enzymes constitution detoxification of the liver (urinary excretion)	Neurological disorders (irritability) Hypermethioninemia Hypouricemia
Selenium (Se)	Fish, nuts, meat, eggs, mushrooms, offal, oysters	50-60 µg	Antioxidant (free radicals) Maintenance of spermatozoa Maintenance of muscles	Neurological delay Osteoarthritis Muscular dystrophy Anemia
Manganese (Mn)	Dried fruits and vegetables, cereals	1-2.5 mg	Antioxidant Carbohydrate and cholesterol metabolism Collagen, bone and joint synthesis	Very rare Excess calcium, Fe or P decreases the absorption of Mn

			Synthesis of sex and thyroid hormones	

2. Vitamins

Vitamins are needed in moderate quantities in the diet. There are 13 vitamins, which are classified into two families:

Water-soluble vitamins:

They are vitamins of the B and C group.

They are soluble in water and eliminated through the urine. They are found preferentially in water-rich foods, such as fruits and vegetables, in cereals (B1), and in meat (B12).

Vitamin B1 (Thiamine)

Vitamin B2 (Riboflavin)

Vitamin B3 (PP or Nicotinic acid or Niacin or Nicotinamide)

Vitamin B5 (Pantothenic acid)

Vitamin B6 (Pyridoxine)

Vitamin B8 (Biotin)

Vitamin B9 (Folates or Folic Acid)

Vitamin B12 (Cobalamins)

Vitamin C (Ascorbic acid)

Fat-soluble vitamins:

Are vitamins A (Retinol), D (Calciferol), E (α-Tocopherol) and K.

They are soluble in lipids.

They can be stored in the body and are not eliminated through the urine.

They are found in oil, meats, butter, colored vegetables.

The following table shows the characteristics of the different vitamins:

Vitamin	Food/Nursery	ANC for adults	Function and role	Pathologies in case of deficiencies
Vitamin A (Retinol)	Carrot, papaya, liver, dairy, sweet potato, palm oil, egg yolk, meat, peach, parsley. coriander;	600-800 µg	Development and growth Night vision Skin health Resistance to infections (immunity) Antioxidant power (hepatic detoxification)	Lung maturation defect Impaired night vision Premature newborn Intestinal and pulmonary infections

Vitamin	Aliments/Food	ANC for adults	Function and role	Pathologies in case of deficiencies
	apricots			Dryness of the skin
Vitamin D (Calciferol)	Fatty fish, egg yolk meats. cod liver oil (Sun=inorganic source)	5 µε	Growth: bone fixation of Ca *[1] Cellular differentiation Immunity (Ca regulation in mammary glands, insulin synthesis, macrophages) Promotes intestinal absorption by ensuring P Ca balance Muscle metabolism (regulation of Ca concentration)	Rickets
Vitamin E (Alpha tocopherol)	Soybeans, corn, sunflower, cereals, nuts	12 mg	Anti-oxidant Muscle and nerve function Stability of the membrane Fertility	Anemia in infants Neurological problems (ataxia)
Vitamin K	egg yolk. poultry, vegetable oil, green vegetables	45 µε	Synthesis of blood coagulation factors	Bleeding problems
Vitamin C (Ascorbic acid)	Orange, tomato, green vegetables, citrus, lemon, green chili, broccoli	110 mg	Anti-fatigue Infection control Iron absorption (red blood cell production) Collagen, adrenaline and amino acid synthesis Antioxidant	Barlow's pathology Scurvy

Vitamin	Aliments/Food	ANC for adults	Function and role	Pathologies in case of deficiencies
Vitamin B1 (Thiamine)	Dried vegetables. nuts. wheat. rice. corn	1.1 - 1.3 mg	Nervous system Carbohydrate metabolism Cardiac function	Wernicke's encephalic disease Beriberi
Vitamin B2 (Riboflavin)/B3 (PP or Niacin)	V. Б2: egg cereals, dairy, liver, green vegetables / V. B3: potato, poultry, lean meat, fish, pulses	1.5mjl 2mg	Cellular Functioning Skin health Oxidation-reduction metabolism	Photosensitive dermatitis Skin lesions Neurological disorders Diarrhea
Vitamin B5 (Pantothenic acid)	Fish, egg yolk, vegetables, cereals,	5mg	Regulation of hormones: Adrenalin, insulin Energy metabolism Regeneration of the skin and mucous membranes	Insomnia Fatigue Gastrointestinal disorders

	mushrooms		Anti stress	Immune deficiency
Vitamin BS (Pyridoxine)	Pulses, cereals, meat, banana, fish, poultry	1.5-1.8 mg	Amino acid metabolism Fetal cell differentiation	Skin disorders Hyperhomocysteinemia
Vitamin BS (Biotin)	Mushrooms, dried vegetables	50 pg	Metabolism of carboxylation + decarboxylation	
Vitamin B9 (Folic acid or Folates)	Raw fruits and vegetables, egg yolk, nuts Almost all foods	300-330 pg	Synthesis of DNA and RNA Anti anemia: red blood cell synthesis Development of the spine and brain in the fetus (3rd month)	Hyperhomocysteinemia Anemia Neurological disorders (ataxia)
Vitamin B12 (Cobalamin)	Milk, cheese, liver, fish, meat, poultry	2.4 pg	Fight against anemia by promoting the absorption of iron in conjunction with vitamin C and the Krebs cycle	Hyperhomocysteinemia Anemia Neurological disorders (sclerosis)

V. Ration or food intake

It is the quantity and nature of food necessary to satisfy the daily needs of an individual in matter (mass) and energy (calories). It must be **quantitatively sufficient** to meet the daily energy expenditure of the subject, and **qualitatively balanced** to ensure optimal intake of amino acids and essential fatty acids, vitamins, minerals and water. The food ration varies according to weight, age, physical activity, sex and height. It is for example estimated at :

AGE	ENERGY REQUIREMENTS (AVERAGE ACTIVITY) KCAL/DAY		ENERGY REQUIREMENTS (AVERAGE ACTIVITY) KCAL/KG	
	♂ ♀	♀	♂ ♀	
Infants				
< 4 months	500	450	94	91
4 to 12 months	700	700	90	θ'
Children and youth				
adolescents	1 100 à 1 300	1 000 à 1 200		
1 to 4 years	1 500	1 400	91	88
437 years old	1 900	1 700	82	78
7 to 10 years old	2 300	2 000	75	68
10 to 13 years old	2 700	2 200	64	55
13 to 15 years old			56	47
Older teens and adults 15 to 19 years	3 100	2 500		
19 to 25 years old	3 000	2 400	46	43
25 to 51 years old	2 000	2 300	41	40
51 to 65 years old	2 500	2 000	39	39
> 65 years old	2300	1 8oo	36	36
			35	35

N.B.: Approximately, the average food intake of a moderately active adult (♂ or ♀) is about 2500 Kcal/d (in Europe) and 2900 Kcal/d (in Morocco: unofficial reference).

For a balanced ration, it is recommended to :

1. Eat 3 main meals a day, with an energy intake of 20% of the TEA at breakfast (morning meal), 40% at lunch (midday meal), 30% at dinner (evening meal), and 10% in the form of snacks (snacks, small snacks, tea, *etc.*), i.e. approximately :

Meals		For the AET of 2500 Kcal/day
Main meals	Breakfast (20% of TEA)	500
	Breakfast (40% of TEA)	1000

	Dinner (30% of TEA)	750
Snack (10% of TEA)		250

2. Provide sufficient daily energy percentages, especially in macronutrients (since minerals and vitamins have no energy value)

Macronutrient distribution

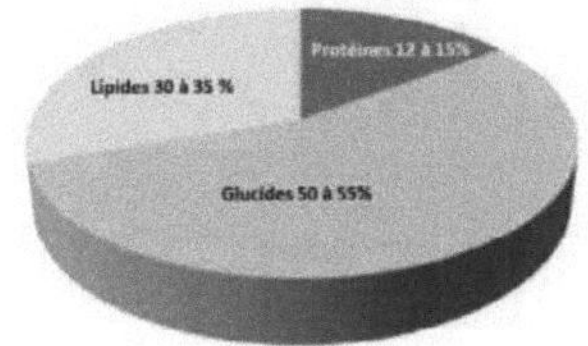

That is, approximately :

Nutrients	Energy intake (%)	For the AET of 2500 Kcal/day
Protein	12 à 15	300 à 375
Lipids	30 à 35	750 à 875
Carbohydrates	50 à 55	1250 à 1375

This is also equivalent to :

Nutrients	Daily intake for 2500 Kcal (in g)
Protein (4 Kcal /g)	75 à 94
Fat (9 Kcal /g)	84 à 97
Carbohydrates (4 Kcal /g)	312 à 344

Nutritional or energy requirements are the amount needed to maintain physiological functions and a normal state of health, and to cope with certain periods of life such as growth, gestation, lactation and old age.

1. Needs of the pregnant woman

The pregnancy phase drastically modifies a woman's physiology, and several adaptations occur in order to store different nutrients in the form of protein, lipid and carbohydrate reserves, in addition to iron and calcium, as soon as the fetus is conceived.

We note:

- A considerable increase in the assimilation of iron, and

- An increase or reduction in urinary waste, mainly of a calcium nature.

Protein requirements are approximately 60g/d

With a supplementation of : 10g/d from $4^{\text{ème}}$ months (in adults) 20g/d from $4^{\text{ème}}$ months (in adolescents)

In carbohydrates

- Favour starchy foods or slow sugars (bread, rice, cereals, legumes, pasta, *etc.*)

- Distribute sugary foods evenly throughout the day

- Do not skip breakfast, especially if you are nauseous

In Lipids: Generally very little changes compared to the normal case, however, the quality of essential GA (*i.e.*: Omega 3) must be regularly satisfied to ensure a good growth of the fetus.

In calcium:

One intake of 1 to 1.2 g/d.

In pregnant women, calcium plays a role in :

- Fetal bone mineralization

- Prevention of high blood pressure

- Calcium enrichment of breast milk

- Possible reduction in the risk of *postpartum* depression (= from the end of childbirth to the first menstrual period after pregnancy)

In Iron:

Iron supplementation of 40 to 60 mg/d in case of anemia due to martial deficiency (= iron deficiency).

In Vitamin B9 :

A supplementation of 0.4mg/d as soon as the contraception is stopped, until the 8 weeks of pregnancy.

In Vitamin D :

A supplementation of 400 IU/d throughout pregnancy, or 100 IU/d in the 3ème trimester, or 100,000 IU as a single dose at 7ème months.

Generally, depending on the traditions and customs of each country, the increase in nutritional needs varies between :

- 100 and 200 Kcal/day, the 1er trimester of pregnancy

- 250 and 350 Kcal/d, the 2ème quarter

Below 1600 Kcal/day, fetal growth may be at risk and the pregnant woman must change her diet and consume daily:

4 dairy products (1000 mg of calcium), 5 fruits or vegetables, 2 portions of protein, 15 to 30 ml of various oils, 1.5 liters of water.

N.B.: In pregnant women, the weight gain should be avoided by not exceeding 12 kg at the end of the pregnancy (at a rate of 1 kg/month).

Certain food hygiene measures (long cooking of meats, good food preservation, use of bleach and disinfectants) must also be respected to avoid contracting **listeriosis** (infection caused by a bacterium of the genus Listeria transmitted by contaminated food, which can cross the placental barrier and reach the fetus), **toxoplasmosis** (infection caused by the unicellular parasite *Toxoplasma gondii*, which lays its eggs in the intestines of cats), as well as food poisoning caused by bacteria such as : Escherichia, Salmonella and Campylobacter. These types of infections can spontaneously interrupt pregnancies, and are responsible for *in utero* deaths of the fetus and certain congenital pathologies of retinal and neurological order.

2. Needs of the breastfeeding woman

For a better production of breast milk, the energy requirements must be increased by :

- 600 Kcal/d, from the 1er day of delivery to the end of the 4ème month

- 500 Kcal/d, beyond 4ème months

- 200 Kcal/d, beyond 4ème months, if breastfeeding is 50% natural and 50% artificial (bottle)

For protein requirements, they must be increased by 15 to 20 g/d.

For lipid and carbohydrate intake, they are similar to those recommended for pregnant women, but for water intake, they are in double quantity compared to the needs of pregnant women in order to allow a good elaboration of the milk (*i.e.*: 3 liters/day of water).

The following diagram illustrates nutrition in the pregnant woman. The same diagram highlights the nutritional needs of the nursing mother:

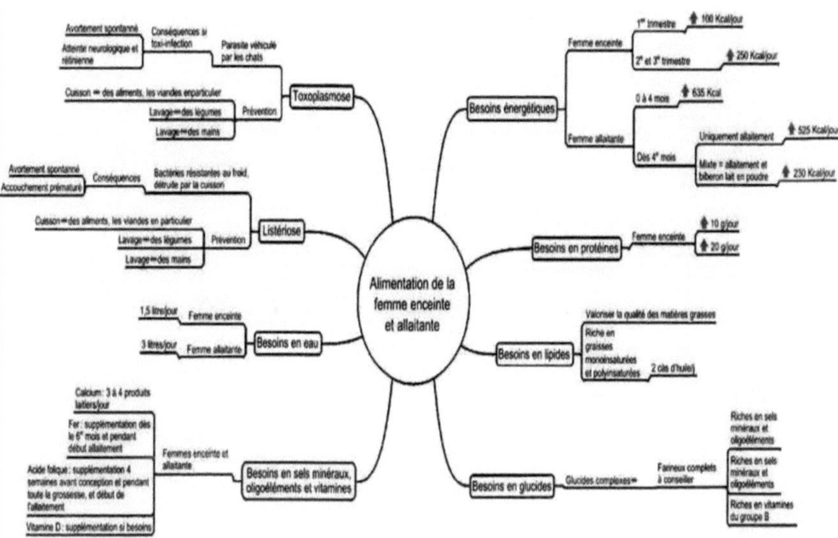

Nutritionists and gynecologists advise breastfeeding women to build up good reserves before delivery, to ensure quality production of breast milk, instead of changing the diet and increasing the food intake after the baby is born.

cf: **The Ministry of Health's 2015 pamphlet on the "National Nutrition Strategy: Program for the Prevention and Control of Micronutrient Deficiencies".**

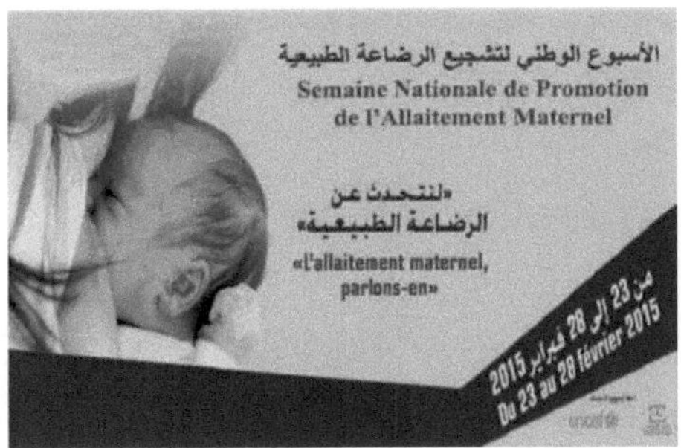

Also, the MOH is holding National Breastfeeding Promotion Week from April 10-16, 2017, under the theme "Let's Help Mothers Breastfeed Their Children, Anywhere, Anytime."

3.   Infant needs

Breast milk is highly recommended by the WHO, as it provides all the nutrients that the newborn needs, especially from birth to 6ème months. Breastfeeding should therefore be exclusive, as it provides infants with proteins, fats, carbohydrates, minerals, vitamins and immunity factors such as antibodies that protect the body against infections and diseases.

However, it is not recommended to give cow's milk to infants, as it is too rich in protein and sodium, poor in essential fatty acids and iron, and can therefore cause digestive problems for babies. Only after 12ème months, a cow's milk intake is suggested, with 1 part water and 2 parts cow's milk. The other alternative that exists is artificial or substitute milk that has a similar composition to the mother's milk.

During the 1er trimester, the newborn's food is based on breast milk + 6 bottles of 10 ml, and from the 4ème month, it increases on average to 4 bottles of 180 ml.

Caloric or energy intake <u>averages 90 Kcal/Kg/d</u>, and is distributed as follows:

From birth to 2 months	97 KcalZKg/d
[3-5 months]	91 KcalZKg/d
[6-8 months]	90 KcalZKg/d
[9-12 months]	96 KcalZKg/d

And in macro and micronutrients, the nutritional intakes are :

Food components	Needs from 0 to 6 months	Needs from 6 to 12 months
Water	150 to 125 ml/kg	125 to 110 ml/kg
Carbohydrates	10-15 g/kg	14-15 g/kg
Protein (represents 15% of body weight)	2-1.8 g/kg	1.5 - 1.4 g/kg
Lipids	4-6 g/kg	
Linoleic acid	3.5-5 % total energy	
Alpha-Linoleic Acid	0.5-5 % total energy	

For minerals and vitamins, they are :

Food components	Needs from 0 to 6 months	Needs from 6 to 12 months
Iron	6-10 mg/d	
Iodine	40 g/d	^{50} µg$^{/j}$
Calcium	400 mg/d	500 mg/d
Vitamin A	350 g/d	
Vitamin B1	0.2 mg/d	
Vitamin B2	0.4 mg/d	
Vitamin B3	3 mg/d	
Vitamin B6	0.3 mg/d	
Vitamin C	50 mg/d	
Vitamin D	20-25 gg/d	
Fluorine	0.25 mg/d	
Phosphorus	100 mg/d	275 mg/d
Magnesium	40 mg/d	75 mg/d
Copper	0.4-0.7 mg/d	0.4-0.7 mg/d
Selenium	^{15} µg$^{/j}$	20 g/d
Zinc	5 mg/d	

Energy	110-100 kcal/kg	95-100 kcal/kg

The transition from exclusive breastfeeding to the consumption of family food is a very delicate phase for infants, and **dietary diversification** (around 6ème months), is necessary to give them safe, adapted complementary foods, and in sufficient quantity so that the transition between breastfeeding and consumption of family food goes well.

The following figure summarizes infant feeding in addition to toddler feeding:

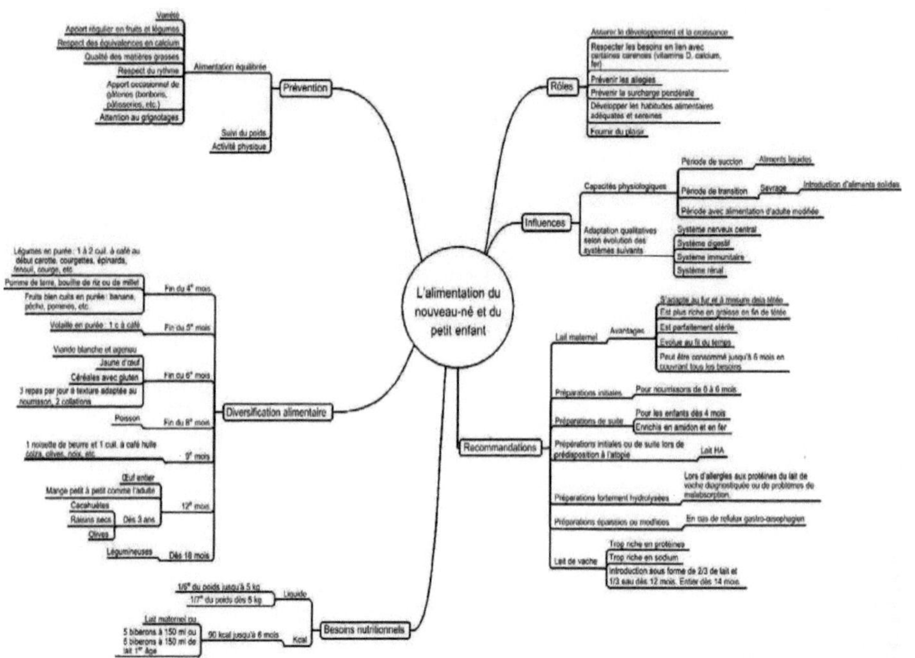

4. <u>Needs of the adolescent</u>

Adolescence is a period of life that extends from puberty (11-12 years) to adulthood (18-19 years). Psychologically, it is a pivotal stage for the development of a unique identity for each person, and for the acquisition of independence and autonomy in order to become an adult.

From a nutritional point of view, the diet must cover the high needs of individuals in full mental development, growth and intense physical activity. In addition, the eating behavior of adolescents is very complex and may follow several norms or trends (*e.g.,* body image: thin for girls and muscular for boys, adherence to an eating style that characterizes a particular social group: musicians, computer scientists, soccer players, models).

The recommended nutritional requirements for an adolescent are shown in the following figure:

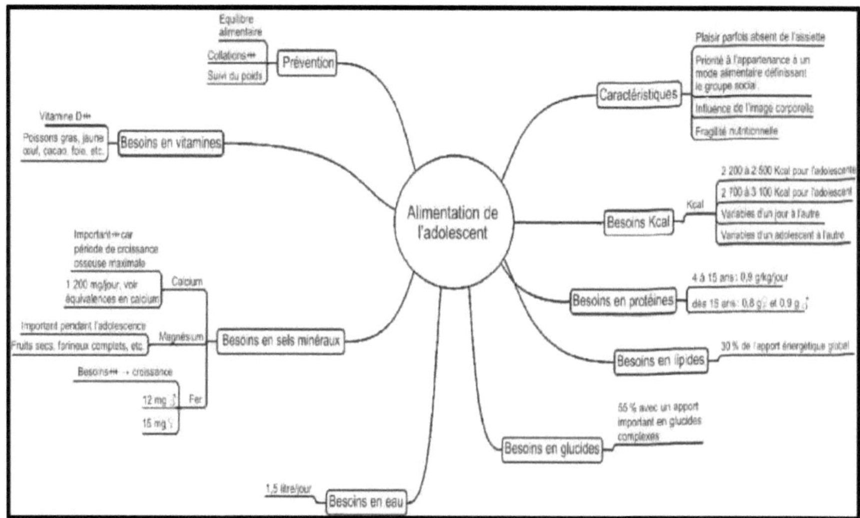

5. Needs of the elderly

Advancing age or aging is often accompanied by a :

- Reduced skeletal or lean muscle mass leading to reduced motor skills

- Loss of autonomy

- Reduced sensation of thirst, hunger, taste and appetite

- Decreased salivary and enzymatic secretion due to psychological problems or medication

- Digestive problems (swallowing and constipation)

The causes of aging are due to a general physiological and metabolic dysfunction of GLP (carbohydrates, lipids, proteins) which is closely related to nutrition. It is therefore crucial to keep the diet of the elderly balanced in terms of energy needs, mainly protein (with an increase of 1g/Kg/d), vitamin needs especially vitamin C and D, water needs to avoid dehydration already progressing with age, and mineral needs mainly calcium to prevent osteoporosis.

In order to maintain skeletal muscle mass in the elderly, maintaining physical activity compatible with aging is highly recommended.

It is therefore important to ensure that the diets of the elderly do not become too monotonous and restricted.

The following figure summarizes the specific nutritional needs of the elderly:

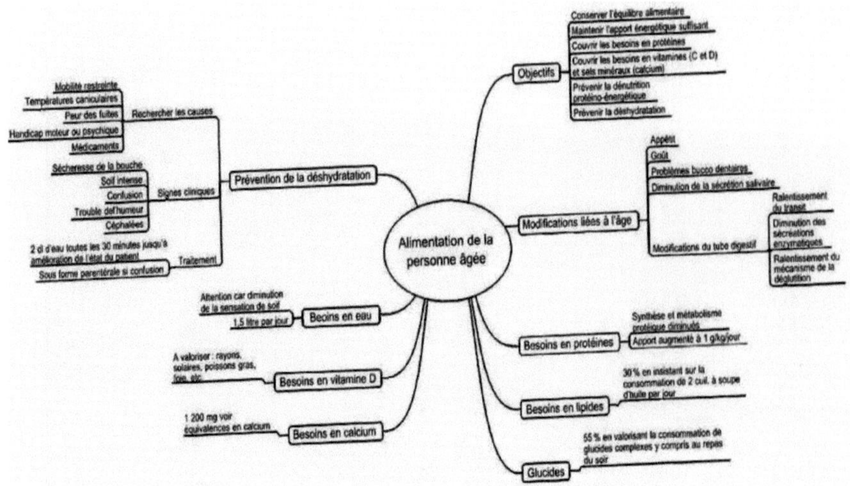

Alimentation de la personne âgée

Objectifs
Conserver l'équilibre alimentaire
Maintenir l'apport énergétique suffisant
Couvrir les besoins en protéines
Couvrir les besoins en vitamines (C et D) et sels minéraux (calcium)
Prévenir la dénutrition protéino-énergétique
Prévenir la déshydratation

Modifications liées à l'âge
Appétit
Goût
Problèmes bucco dentaires
Diminution de la sécrétion salivaire
Modifications du tube digestif
Ralentissement du transit
Diminution des sécrétions enzymatiques
Ralentissement du mécanisme de la déglutition

Besoins en protéines
Synthèse et métabolisme protéique diminués
Apport augmenté à 1 g/kg/jour

Besoins en lipides
30 % en insistant sur la consommation de 2 cuil. à soupe d'huile par jour

Glucides
55 % en valorisant la consommation de glucides complexes y compris au repas du soir

Prévention de la déshydratation
Rechercher les causes
Mobilité restreinte
Températures caniculaires
Peur des fuites
Handicap moteur ou psychique
Médicaments
Signes cliniques
Sécheresse de la bouche
Soif intense
Confusion
Trouble de l'humeur
Céphalées
Traitement
2 dl d'eau toutes les 30 minutes jusqu'à amélioration de l'état du patient
Sous forme parentérale si confusion

Besoins en eau
Attention car diminution de la sensation de soif
1,5 litre par jour

Besoins en vitamine D
À valoriser : rayons solaires, poissons gras, foie, etc.

Besoins en calcium
1 200 mg voir équivalences en calcium

1. Importance of food hygiene

Careful food hygiene reduces the risk of contamination and food poisoning, and consequently, the danger of intestinal discomfort and digestive disorders.

According to the report of the anti-poison and pharmacovigilance center of Morocco (CAMP), a total of 2670 food poisoning were reported in 2013. Food poisoning is therefore a matter of concern for public health in our country.

The causes of food poisoning are frequently related to the poor control of hygiene and sanitation rules from the phase of cultivation, breeding, production, processing, distribution of foodstuffs, storage and purchase, to the phase of preparation and consumption of food.

Symptoms of food-related illnesses are:

1. Diarrhea (dehydration due to loss of water and electrolytes such as sodium and potassium)

2. Vomiting

3. Nausea

4. Abdominal cramps

5. Constipation

6. Persistent fever

7. Headaches

Food poisoning due to ***Listeria*** bacteria can cause problems for the unborn child, while an ***Escherichia coli*** infection can cause kidney problems. Other complications such as arthritis and bleeding problems may occur. Non-infectious food poisoning can sometimes lead to permanent nervous system problems and even death.

Safety tips and good hygiene practices

In order to avoid these nuisances, it is important to keep food safety to a minimum and to take the following precautions:

S Constant hand washing (drinking water, soap, gel), especially after changing a baby, leaving the toilet, handling raw meat, poultry, fish or seafood, or after contact with pets.

S Avoid contact between raw and cooked foods, e.g. raw meat and raw food.

S Washing foods such as fruits and vegetables, fish, meat, poultry and eggs with potable water before cooking or eating.

S Personal hygiene in the kitchen (clean clothes and aprons, hair tied back, clean fingernails, clean bandages in case of injuries or cuts. Avoid coughing or sneezing on dishes in case of illness or flu!)

S Maintenance and cleaning after each use, of kitchens, surfaces and workplaces using disinfectants, bleach (1 ml of bleach in 150 ml of water), sponges, garbage bags, gloves, detergents, descalers, cloths, *etc*.

S Maintenance, cleaning and temperature control for the cooler at 0°C, the refrigerator at 4°C and the freezer at -18°C in order to prevent the growth and propagation of bacteria, parasites or viruses on the food.

S Food should be packaged and stored carefully and within the recommended storage temperatures or capacities for each type of food and for each situation: during food purchase, cooking, freezing, refrigeration, thawing, *etc*.

During the purchase of food

It is crucial to check:

- The use-by date on the products purchased is marked "use by". Once passed, the products are to be thrown away because they are unfit for consumption.

- Best before dates are expressed as "best before" for groceries, frozen foods and beverages. When this date is exceeded, the product can be consumed without risk, but it has lost some of its nutritional and organoleptic qualities.

- The storage temperatures indicated on the packaging.

- Frozen and deep-frozen products should be purchased last, put in insulated bags and placed quickly in the freezer or prepared immediately.

Cooking

Food must be carefully cooked to a high core temperature: simply reheating the food is not enough to eliminate pathogens. Be careful, reheating in a microwave oven does not ensure a homogeneous temperature for the whole food.

2. Qualities of a good food

A good quality food must include **the 4S**: Health, Safety, Taste and Service.

- ***Health or nutritional quality***: depends on its concentration in micronutrients and its energy intake.

Foods with good nutritional quality have few calories but a large amount of micronutrients (minerals and vitamins), for example: fruits and vegetables, fish, grain products and some dairy products. These types of foods are very healthy and ensure a balanced diet.

Foods of low nutritional quality therefore contain few micronutrients and many calories, for example: pastries, breaded foods, sweets, pizzas.

- *Safety or hygienic quality*: The food must be healthy and respect the rules of hygiene and salubrity in order to avoid food poisoning and contamination (see section: food hygiene).

- *Taste or organoleptic quality (= gustatory = hedonic)*: a quality food must also satisfy the 5 senses of the organism (taste, hearing, vision, smell and touch). It is said that organoleptic quality has **a major sensory component**.

The gustatory quality has also **a psychological component** (Perception or the biophysical, physio-psychological and cultural faculty that connects the action of the living to the environment *via* physiological senses and individual or collective ideologies) and **a socio-cultural component** since the circulation of the food indicates the social relations between the individuals and assure the transmission of the culinary, cultural and other habits inside the communities).

- *Service or quality of storage, preservation, use and acquisition*: For the consumer, a good food must have the following qualities: easy to find (available and on sale everywhere), easy to prepare (canned or frozen products), easy to store for a long time after their purchase (honey, UHT milk, dry pasta, chocolate), and a reasonable and affordable quality/price ratio.

3. <u>Food preservation</u>

Is done by cold (refrigeration and freezing) or following traditional methods of conservation.

a. *Refrigeration and freezing*

The recommended refrigeration times are for safety and the freezing times are for quality. It's simple, the longer you refrigerate packaged foods in the freezer, the longer the quality of the food is preserved.

In a refrigerator, the temperature is often not uniform inside. Always make sure that the coldest part is at the top or at the bottom. The following diagram shows the temperature distribution on the different levels of a refrigerator and tips for a better preservation of different foods:

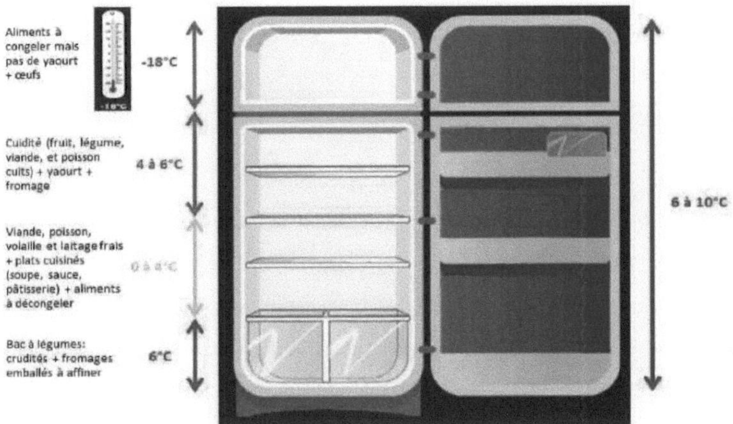

Refrigeration or freezing does not destroy microorganisms, but rather reduces or slows down their growth. Only a temperature > 63°C allows this destruction. Stopping the cold chain can lead to new proliferation of micro-organisms which can become pathogenic in some cases.

The cold chain must be respected especially in extreme cases where no refrigeration or freezing equipment is available (*e.g.* during a trip, a desert excursion, a picnic). In this case, the use of coolers, ice packs, or isothermal bags can slow down the sudden increase in temperature, and thus limit the possible growth of microorganisms (however, for a limited time).

<u>Food storage in the refrigerator and freezer</u>

Food	Refrigerator at 4°C minimum	Freezer at - 18 °C minimum
Fresh meat		
Beef	From 2 to 4 days	From 10 to 12 months
Lamb	From 2 to 4 days	From 8 to 12 months
Minced meat	From 1 to 2 days	From 2 to 3 months
Fresh poultry		
Whole chicken/turkey	From 2 to 3 days	1 year
Chicken/Turkey - pieces	From 2 to 3 days	6 months
Fresh fish		
Lean fish - cod, sole, *etc.*	From 3 to 4 days	6 months
Fatty fish - salmon, etc.	From 3 to 4 days	2 months
Mollusks and crustaceans	From 12 to 24 hours	From 2 to 4 months
Scallops, shrimp, cooked mollusks	From 1 to 2 days	From 2 to 4 months
Bacon and sausage		
Bacon	1 week	1 month
Raw sausages	From 1 to 2 days	From 1 to 2 months

Table scraps		
Cooked meat, stew, eggs or vegetarian dishes	From 3 to 4 days	From 2 to 3 months
Poultry and cooked fish	From 3 to 4 days	From 4 to 6 months
Broth and sauce	From 3 to 4 days	From 4 to 6 months
Soups	From 2 to 3 days	4 months
Frozen meals		
Keep frozen until ready to cook		From 3 to 4 months
Eggs		
Fresh eggs with shell	From 3 to 4 weeks	Do not freeze
Fresh eggs without shells	From 2 to 4 days	4 months
Hard-boiled eggs	1 week	Does not freeze well
Dairy products		
Unopened milk carton	Expiration date	6 weeks
Open milk carton	3 days	Do not freeze
Unopened cottage cheese container	Expiration date	Does not freeze well
Open cottage cheese container	3 days	Do not freeze
Unopened yogurt container	Expiration date	From 1 to 2 months
Open yogurt container	3 days	Do not freeze
Soft cheese	1 week	Does not freeze well
Semi-firm cheese	From 2 to 3 weeks	8 weeks
Firm cheese	5 weeks	3 months
Hard cheese	10 months	1 year
Processed cheese	5 months	3 months
Unwrapped salted butter	8 weeks	1 year
Unwrapped unsalted butter	8 weeks	3 months
Unwrapped butter	3 weeks	Do not freeze
Vegetables		
Green or yellow beans	5 days	8 months
Carrots	2 weeks	From 10 to 12 months
Celery	2 weeks	From 10 to 12 months
Leaf Lettuce	From 3 to 7 days	Do not freeze
Spinach	From 2 to 4 weeks	From 10 to 12 months
Zucchinis	1 week	From 10 to 12 months
Tomatoes	Do not refrigerate	2 months

Source: Government of Canada

b. *Traditional food preservation*

► Oven or sun <u>drying</u> by evaporation

► <u>Salting or curing,</u> which consists of preserving food by using salt

► <u>Brining and pickling</u> using vinegar

▶ <u>Pasteurization</u> by heating at a temperature between 60 and 90°C to eliminate microorganisms

▶ <u>Sterilization</u> by heating to a temperature of 120°C

4. Defrosting

It is important to follow the thawing instructions on the product label and never refreeze a product that has already been thawed.

Food consumption is an extremely complex individual and social phenomenon. It responds to a multitude of interacting internal and external factors, including food availability, dietary diversity, lifestyle, and socio-economic level (FAO, 1997). The meal plays an essential social role and the mode of consumption is indeed characterized by a certain conviviality around the structured meal obeying a certain ritual and respect for the food.

In Morocco, the diet and habits are of the **Mediterranean** type (= olive oil, olives, cereals, herbs/spices, fruits/vegetables, nuts, fish/seafood, legumes and alcoholic beverages), and are mainly characterized by a large consumption of cereals, fruits and vegetables.

of energy intake by food groups 2005-2007

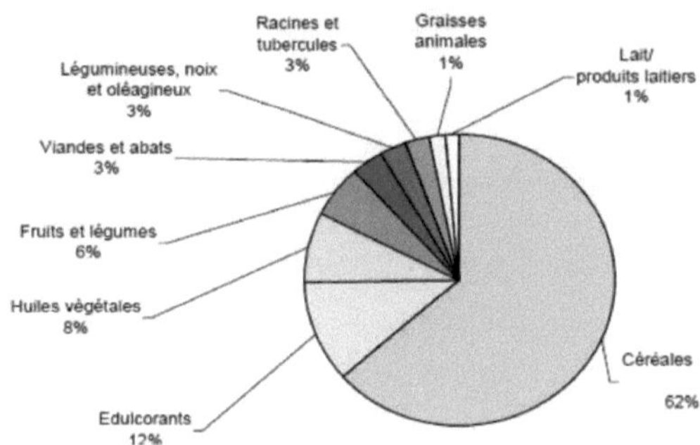

Maroc
Source: FAOSTAT
Note : les groupes <1% (poissons et fruits de mer, oeufs et autres) ne sont pas présentés dans ce graphique.

Even though the diet consists of so many micronutrient-rich foods, the consumption of animal products remains modest, even though the country's animal and fishery resources are immense.

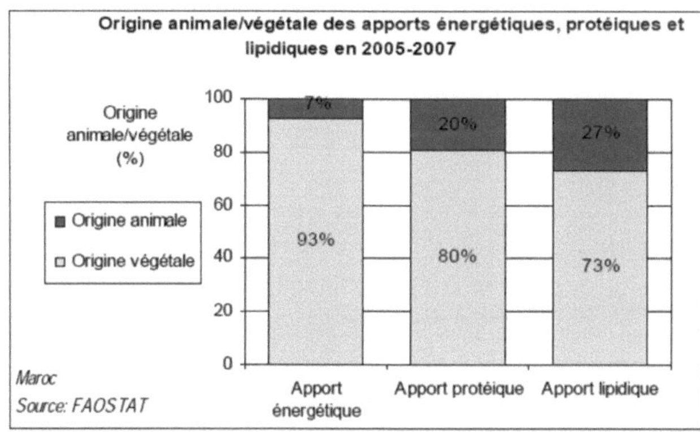

At the same time, Morocco is experiencing a food transition that affects both urban and rural areas.

Urbanization, economic development and globalization are causing changes in eating habits.

Eating out (fast food) is also becoming more common, especially in urban areas, which encourages the consumption of foods that are higher in sugar and fat.

The evolution of energy intake by macronutrient (GLP)

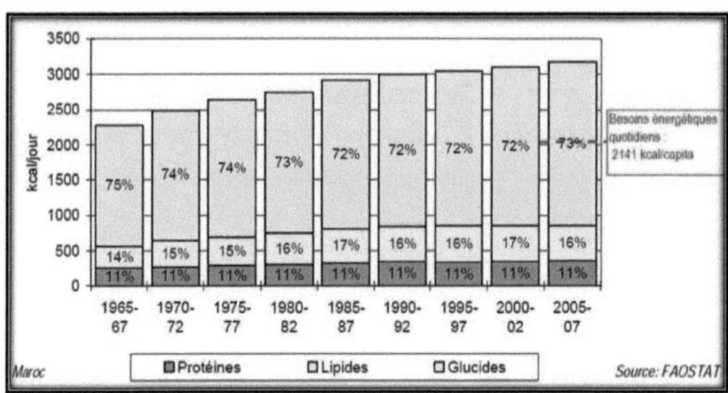

In addition to this, there is generally a reduction in physical activity (sedentary mode) among Moroccans, which is at the origin of the progression of overweight and obesity in the adult population (*Nutritional Profile of Morocco - Nutrition and Consumer Protection Division, FAO, 2011*).

However, the traditional Mediterranean diet, recognized for its health benefits, is still favored in Morocco. Indeed, there is a return to **traditional and local food products** (= local organic products *vs.* GMO: olive and argan oils, whole grains, barley, cactus fruit, mallow, saffron...*etc.*), as well as to traditional culinary methods (home preparation of bread or couscous).

Different aspects of Moroccan cuisine and eating habits

Moroccan cuisine is an exotic cuisine rich in tropical spices, vegetables, seafood, and is incredibly delicious. Moroccan eating habits are influenced by a multitude of cultures, especially Arab, Berber, Andalusian, African and Mediterranean.

Moroccans eat 3 meals a day in addition to the snack, and take great care as to the freshness and quality of products. Everything that is produced in the country or "beldi" product, is very popular in Morocco (Jben, Smen, honey, eggs and chickens beldi.*etc.*).

The breakfast

A very hearty morning meal, with bread, cheese, pancakes, baghrir, msammen, olive oil, Argan oil, Amlou, mint tea, Asekkif or dchicha soup, black or milk coffee.

The lunch

A typical lunch meal begins with a series of hot and/or cold salads, followed by a tagine, couscous or dwaz, and ending with a cup of mint tea or fresh fruit for dessert. Moroccans use the fork, knife and spoon, but prefer to eat with their hands using bread as a utensil. The consumption of pork and alcohol is not common for religious reasons. The concept of bleeding meat does not exist, and this is for hygienic purposes and especially to respect the religious Halal mode of slaughter.

The dinner

Soups (vegetable, dchicha, semolina)

Grilled fish or meat

Harira

International pasta: Spaghetti, Lasagna

International dishes: Paella, Pizza

Dairy with bread (yogurt, cheese, lben... *etc.*).

Raw fruit/fruit salad

Dessert (Flan, Moroccan pastries)

Miscellaneous :

Examples of dishes served at snack time, at Ftour/S'hour of Ramadan, and at breakfast:

Harcha

Krachel

Crescent

Chebbakia

Baghrir

Msemmen

Traditional cake

Rayeb

Khliaa with eggs

Rghayef

Thride

Pancake

Stuffed or not stuffed Batbot

Briouat with almonds or meat, fish, spinach, *etc.*

Khobz (all varieties of bread: tafarnout, wholemeal, baguette, half, rusk, etc.)

<u>Examples of dishes eaten for lunch and/or dinner:</u>

Brochette (kabab or boulfaf) of meat, chicken, fish

Couscous

Merguez (meat sausage: mutton/beef/turkey...)

Quaddid

Stuffed chicken

Kefta or minced meat

Lentil Soup

Pea soup or Bissara

Mechoui (grilled meat)

Mrouzia

Pastilla

Rfissa

Tanjia

Tagine (of meat, chicken, fish or vegetables)

Fish Frying

Spicy and cooked mussels

Hrira

Vegetable soup

Baddaz

<u>Condiments (for seasoning), sauces and salads</u>

Harissa (made of garlic, chillies, coriander and cumin, ground together)

Limes (pickled lemons)

Marinated olives

Charmoula (marinade)

Zaalouk

Chakchouka

Mixed cooked vegetable salad

Mixed raw fruit salad

Pickle

<u>Traditional desserts and pastries</u>

Ghraiba

Almond brioche

Baklawa

Fekkas

Kounafa

Kaab Ghzal

Chopped carrots and orange juice with cinnamon

Seffa

Sellou

Doughnut (Sfanj)

Rice with milk

<u>Drinks</u>

Mint tea (Atay b naânaâ)

Tea without mint

Wormwood tea (Atay b chiba)

Fruit juices (orange, lemon, grenadine, grapes... *etc.*)

Coffee with milk

Drinking water/mineral. *etc.*

Orange blossom water (Ma Zhar)

Rose water (Maa Louard)

Licorice (Aarkssouss)

<u>Spices</u>

Zaâter (Oregano)

Felfla soudania (Dry pepper)

Ourkat sidna moussa (Laurel)

Habbat al hil (Cardamom)

Karfa (Cinnamon)

Kamoun (Cumin)

Tahmira/felfla hemra (Paprika or red pepper)

Jenjlane (Sesame seed)

Kesbour (Coriander)

Zaafran beldi (Saffron: the most expensive spice in the world)

Kronfel (Clove)

Basbas (Fennel)

Kharqoum (Turmeric)

Skinjbir (Ginger)

Lbzar (Black pepper)

Nafaâ (Anis)

Meska El Hourra (Gum Arabic)

Elgouza (Nutmeg)

Maadnouss (Parsley)

Krafess (Celery)

El Halba (Fenugreek)

Ras El Hanout (blend of 37 spices)

<u>Herbs and medicinal plants</u>

Naanaa (Mint)

Fliyo (Peppermint)

Merdedouch (Marjoram or Garden Oregano)

Kerouiya (Caraway or meadow cumin)

Ellouiza (Verbena)

Kebbar (Caper)

Hbak or Rihane (Basil)

Salmiya (Sage)

Khzama (Lavender)

Avitaminoses are vitamin deficiencies. Generally, in industrialized countries, nutritional and vitamin deficiencies are correlated with unbalanced nutrition, chronic digestive malabsorption, pregnancy, old age and chronic alcoholism.

In poor or developing countries, deficiencies are the result of famine, malnutrition and wars.

1. Avitaminosis A

Definition	Possible causes	Symptoms	Treatment
Vitamin A deficiency	Intestinal malabsorption Diabetes Ulcerative Colitis Celiac disease Cystic fibrosis Hypothyroidism Cirrhosis of the liver	Eye problems: at night, dry eyes, infantile blindness due to corneal damage Dry skin Growth retardation in children	Long-term intake (1 to 3 months) of vitamin A tablets Food rich in vitamin A

2. Avitaminosis B

Definition	Possible causes	Symptoms	Treatment
Vitamin B deficiency (B1, B2, B6, B9 and B12)	Pregnancy Old age Digestive malabsorptions Chronicles Chromic alcoholism Intestinal surgery Dietary imbalance: * vegetarian diet (B12), * no-residue diet (B9), * weight loss diet (B1, B3, B6) Association of a deficiency in vitamin B1 with another in	Beriberi (B1 deficiency) Problems: skin, muscle, eye (c. in B2) Irritability, depression, loss of appetite, weight loss, anemia (c. in B6) Crohn's disease = complex chronic inflammatory disease of the digestive tract (c. in B9 or folates) Celiac disease = autoimmune disease caused by intolerance to gluten (cereals) and	Tablets for vitamin B1, B2 and B6 Intravenous for vitamin B12 No drug treatment for celiac disease Taking 0.4 mg/day of vitamin B9 before conception and during the first 3 months of pregnancy reduces the risk of spina bifida

	vitamin B6 and B12	characterized by the destruction of the villi of the small intestine, and consequently, malabsorption of nutrients, such as iron, calcium and folic acid (c. in B9)	
		<u>Spina bifida</u> = neural tube defects and congenital malformation of the spine (c. in B9) <u>Biermer's disease</u> (c. in B12)	

3. <u>Avitaminosis C</u>

Definition	Possible causes	Symptoms	Treatment
Ascorbic acid or vitamin C deficiency	Malnutrition and starvation Digestive absorption problems Chronic alcoholism (in the elderly)	<u>Early Avitaminosis, usually asymptomatic:</u> aches and pains for minimal effort, muscle and/or bone pain, irritability <u>Scurvy:</u> tooth loosening, gum purulence, edema, hemorrhages, anemia and signs severe neurological damage (coma) in very advanced forms and then death	Diet rich in fruits and vegetables Tablets (1g) every morning Vitamin C injection

4. <u>Avitaminosis D</u>

Definition	Possible causes	Symptoms	Treatment
Vitamin D deficiency	Genetic pathologies (cystic fibrosis) Chronic digestive malabsorptions Liver and pancreas pathologies Stomach diseases or surgery Kidney pathologies Hypothyroidism Chronic alcoholism Medication	<u>Rickets</u> in children due to insufficient mineralization and calcification of bones and cartilage, or lack of magnesium, calcium and vitamin D. Rickets is also characterized by **bone deformations,** the appearance of swellings (lumps) in the extremities of the long bones, diffuse bone and muscle pain	<u>Preventive treatment:</u> 400 IU/d of vitamin D for children exposed to the sun 800-1200 IU/d for children with little sun exposure and pigmented skin = (dark skin filters UV rays more effectively, which limits skin cancers but also limits vitamin D synthesis.
		<u>Deficiency osteomalacia</u> = bone decalcification in adults with extreme difficulty walking, muscle fatigue and bone pain (*e.g.* chronic low back pain, knee pain).	The light skin of populations living far from the equator, in regions of low sunlight, would be the result of a natural evolution to avoid rickets) <u>Curative treatment:</u> Taking vitamin D + calcium

5. Avitaminosis K

Definition	Possible causes	Symptoms	Treatment
Vitamin K deficiency, which plays a role in blood	Intestinal malabsorption After surgery to remove part of	<u>Hemorrhages, anemias:</u> Gingivorrhagia (bleeding or hemorrhage from the gums)	Oral vitamin K intake

| clotting | the small intestine Ulcerative Colitis Crohn's disease Biliary retention Celiac Disease Taking anti-vitamin K drugs: anticoagulants, antibiotics | Epistaxis (nosebleed) Rectorrhagia (blood in the stool) Hematuria (blood in the urine) Breakthrough bleeding (bleeding from the vagina) Anemia in severe or chronic deficiencies | |

SECTION 2: DIETS

Thanks to a healthy and balanced nutrition and diet, the appearance of many pathologies and affections can be prevented. However, if nutrition can prevent certain health problems, it must be adapted and applied in case of illness.

Certainly, dietetics, through its therapeutic applications, has become an important component in the treatment of certain pathologies that require special diets, either reduction or exclusion (of certain foods).

By definition, a diet is a modification of an individual's spontaneous diet with the aim of eliminating momentarily or permanently harmful effects.

Medical or therapeutic diets are dictated in cases of illness by a doctor or dietician who determines their duration and methods of application.

This prescription requires a diagnosis (diabetes, hyperlipidemia, renal insufficiency, *etc.*), proof that the diet is effective for the condition in question, and that there are no contraindications or side effects. The objective of the diet and its modalities must then be explained in detail (authorized foods, prohibited foods, *etc.*), the patient must be convinced, his understanding must be ensured, and sometimes the diet must be adapted to his habits, so that he can adhere to it.

A. Reduction schemes

This type of diet aims to reduce and limit the quantities or energy intake of one or more foods (without eliminating them permanently).

Diet: Normal light or digestive sparing :

It is prescribed in order to limit flatulence (production of intestinal gas), digestion time or during digestive discomfort in relation to :

- Post-operative re-feeding

- Heart failure

- Inpatient Plan

- Dyspepsia and digestive disorders (irritable bowel)

- Anxiety and stress

- Drug-induced gastritis

- Increased motility of the digestive tract

- Physical activity

- Elderly people with digestive difficulties

- Asthenia

- Pregnant woman in late pregnancy

Permitted foods:

Steam cooking,

Grilled chicken without skin,

Marbled mushrooms,

Parsleyed zucchini,

Rice salad,

Paella,

Vegetable flan,

Baked apple,

Apricot jam,

Pear pie,

Milk semolina,

Dry cookies,

Peaches in syrup,

Cereals

Prohibited foods:

Hyperlipidic foods,

Food with a strong taste and causing the production of gas,

Foods that delay gastric emptying and trigger the sensation of gastric heaviness, and generate burning sensations

Anti-diarrhea diet

It is dictated to people with diarrhea, in order to avoid dehydration and to slow down the digestive transit.

Permitted foods:

Lots of H2O,

Meat, fish and eggs with simple cooking such as grilling, boiling, without sauce, or with a meat juice,

Starchy foods (potatoes, pasta, rice, semolina, wheat) plain, in juice or with a little butter,

Cooked carrots,

Fruits: Banana, baked apple, applesauce and apple-banana,

Lactose-free dairy products

Prohibited foods:

Raw and cooked green vegetables

Dry vegetables

Raw fruits and juices

Regular milk

Sauces, fries and fats

Soft drinks

Colic/constipation saving diet

It is intended for people with colopathy or digestive disorders such as colitis (inflammation of the colon), constipation, or post-operative disorders in digestive surgery. The goal is to limit digestive disorders and facilitate the return of transit.

a.   In the case of *colonic sparing* (= resting the colon), all transit gas pedals constituting sources of lignin, cellulose, and poorly digestible starch must be eliminated.

Permitted foods:

Tender meats, fish and eggs with simple cooking such as grilling, boiling, without sauce or with a meat juice,

Low-fiber vegetables (zucchini, cooked lettuce, carrots, asparagus tips...),

Starchy foods (pasta, rice, potatoes, semolina, wheat) plain and without sauce,

Low-fat dairy products,

Cooked or very ripe fruits,

Hard cheeses (Gruyere, Edam),

Moderately sweetened beverages

Prohibited foods:

Dry vegetables,

Vegetables that promote flatulence: peas, salsify, artichokes, mushrooms, all cabbages (flower, broccoli, red, etc.),

Bread and pastries,

Garlic, onions, shallots,

Cooked fats, cold cuts, sauces and fried foods,

Crudités to be tested according to tolerance,

Spices and condiments,

Milk and dairy products (in case of lactose intolerance)

b.   *Constipation* = slowing down of the intestinal transit, generating a decrease in the frequency of stool emission, which will appear dehydrated. It is often caused by a diet deficient in dietary fiber, which is full of water and facilitates food transit.

Permitted foods:

Water,

Legumes,

Vegetables rich in fiber: celery, spinach, fennel,

Whole grains: brown rice, wheat,

Flaxseed,

Dried fruits: especially prunes

Fresh fruit: plums, raspberries, blackberries,

Oat bran

Prohibited foods:

Cabbage, radishes, refined rice, potatoes, bananas,

Spicy dishes,

Alcohol,

Fermented cheese,

Fried and fatty foods

N.B.: To fight constipation, you can use laxatives or probiotics.

Laxatives

They are made up of insoluble dietary fibers and soluble mucilage that swell when in contact with water, making the stool more dense and voluminous (hydrated), which promotes peristalsis *via* acidic fermentations (starch, digestive cellulose) and then facilitates their progression through the intestine. They are of 2 types:

- Natural laxatives like flaxseed (zerriaat al kettane) and oat bran (noukhala)

- Pharmaceutical laxatives.

Probiotics

These are intestinal bacteria (intestinal flora) that facilitate transit and therefore digestion. Among the probiotics found in most dairy products are Lactobacillus, Bifidobacterium, and *Sacchromyces boulardi*. The use of probiotics is still debated because of its side effects.

Diet : Standard diabetic

Reminder on the different types of diabetes mellitus :

There are at least 4 main types of diabetes:

- Type 1 diabetes= (insulin dependent diabetes = IDDM) = lean = juvenile

- Type 2 diabetes* = (non-insulin dependent diabetes = NIDDM) = fat

- Gestational diabetes

- Pre-diabetes

- In the standard diabetic diet for people with type 2 diabetes or pre-diabetes, it is important to :

> Consume approximately 120 to 240 g of carbohydrates per day

> Split meals (5/d)

> Reduce fats

> Eat at fixed times

Permitted foods:

Lean meat, fish, eggs, poultry, H2O

Food to be measured:

Cereals, rice, pasta, oil and butter.

Less sweet vegetables:

Lettuce, spinach, green beans, leeks, zucchini, artichokes, mushrooms.

Sweeter vegetables:

White beans, lentils, peas, carrots, salsify, beets, potatoes.

Less sweet fruits:

Watermelon, orange, melon, grapefruit.

Sweeter fruits:

Banana, apple, grape, pear, peach. Bread and dairy products

Prohibited foods:

Meat in sauce,

Industrial charcuterie,

Cooked fats,

Industrial soups,

Dried fruits,

Starches and oilseeds,

Syrups,

Pastries,

Honey,

Chocolate,

Ice cream,

Alcohol,

Industrial fruit juices

Diet : Hypocaloric (limiting calories and energy)

It is indicated in case of :

- Overweight or obese,

- Hypertriglyceridemia

- Type 2 diabetes

It is based on a significant reduction in energy intake in order to force the body to draw on its reserves. This diet requires a very rigorous and long term follow-up to obtain results (the total energy intake must be about 2/3 of the initial caloric intake to maintain a 10% weight loss).

Permitted foods:

Lean meat and fish,

Green vegetables and fruits,

Low-fat and low-sugar products

Prohibited foods:

Fast-absorbing sweet foods: honey, chocolate, candy, pastries, sodas, creams,

Fats, deli meats

Alcoholic beverages

Starches

N.B.: Side effects to weight loss may have occurred:

- Uncontrolled progressive diseases,

- Severe depressive states,

- Severe psychiatric illnesses,

- Severe eating disorders,

Diet: Hypolipid (limiting lipids and fats)

It is prescribed in case of :

- Biliary or hepatic or pancreatic insufficiency,

- Arteriosclerosis

Permitted foods:

Lean meat and fish,

Cheese at 0%,

Raw vegetable fats,

Fresh fruits and green vegetables

Prohibited foods:

Fatty meats and fish,

Eggs,

Dairy products,

Alcohol,

Sugars,

Dried fruits and vegetables,

Cooked fats

Cholesterol-lowering diet

It is prescribed to people who have a very high level of LDL cholesterol in their blood, which can cause blood vessels to become clogged and thus increase the risk of cardiovascular disease. It is possible to lower this level by modifying and controlling your diet.

Foods allowed or to be limited:

Increase mono- and poly-unsaturated oils: olive, corn, walnut, sunflower,

Increase fiber: whole grain bread, salad,

Encourage daily consumption of fruits and vegetables,

Encourage the consumption of fish 3 times a week,

Encourage the consumption of starchy foods and pulses

Limit the intake of saturated fatty acids: butter, eggs, cheese, cold cuts,

Reduce alcohol intake and consume plenty of H2O

Prohibited foods:

Offal,

Sugars,

Alcoholic beverages,

Egg yolk,

Sauces and fried food,

Whole milk and all dairy products made from whole milk,

Fatty meats: mutton, lamb (except leg), poultry with skin, game, sausages, merguez, Mollusks, mainly oysters,

Oleaginous fruits: walnuts, hazelnuts, almonds,

Chocolate

Anti-triglyceride or hypo-triglyceride diet

It is dictated to people with too high a blood concentration of triglycerides, or hypertriglyceridemia. Triglyceridemia is affected by various factors, such as age, weight, diet, smoking, alcohol, physical activity, certain oral contraceptives containing estrogen and certain medications, such as benzodiazepines and corticosteroids.

Thus, hypertriglyceridemia often occurs with risk factors such as abdominal obesity, high blood pressure, high blood glucose and low HDL cholesterol.

Permitted foods:

Sugar-free desserts (sweetener allowed),

Sugar-free dairy products (sweetener allowed), limited to 3/day,

Fresh or stewed fruit with no added sugar (limited to 1 or 2 per day)

Foods rich in Omega-3: fish,

Cereals: rice, barley, whole wheat,

Bread

Prohibited foods containing simple sugars:

Flavored or fruit (fructose) milk products,

Sweetened condensed milk (lactose),

Entremets, ice creams, and sorbets,

Dried fruits, candied or in syrup,

Honey, chocolate,

Cakes, cookies, pastries,

Sodas, syrups, fruit juices,

Alcohols: wines, aperitifs, beer

Diet : Hypo-protein (reducing proteins)

It is prescribed for people with renal failure who are not on dialysis*.

Permitted foods:

Patients should take between 40 and 60g of protein/day, and a diet containing :

Meat, fish and eggs in moderation.

Vegetables, starchy foods, fruits, bread

Prohibited foods: (salty foods) :

Cheeses, canned, smoked and salted meats,

Charcuterie

Canned, smoked, dried and salted fish,

Shellfish and crustaceans,

Freeze-dried products,

Sodas,

Mineral and carbonated waters

Salty condiments: mustard, pickles, celery salt, olives

Dry vegetables

Dried fruits and oilseeds

Dialysis diet

It is intended for people undergoing dialysis. It is a salt-free diet, with water restriction (quantity of liquid/day authorized).

It is a diet rich in proteins but limited in potassium contained in fruits and vegetables in large quantities.

Permitted foods:

Salt-free foods

Proteins such as meat, fish, eggs, dairy products,

Fruits and vegetables combined, we count 1 raw and 2 cooked per day,

Starchy foods and bread

Prohibited foods:

Salty foods: Cheese,

Canned, smoked or salted meats, sausages and hams,

Canned, smoked, dried or salted fish,

Shellfish and crustaceans,

Industrial dishes including pastries, freeze-dried soups, freeze-dried coffee,

Sodas and flat or carbonated salted mineral water

Savoury condiments

Dry vegetables

Dried fruits and oilseeds

Banana and chocolate, cream, fruit in syrup

Soups and broths (due to water restriction)

B. Exclusion, avoidance or "no" diets

This type of diet is based on the exclusion or avoidance of the food responsible for a specific pathology or allergy (allergen removal). The prescription of the diet requires a precise and definitive diagnosis of a specific pathology.

Diet : San sel

It is prescribed in the case of a :

- heart or kidney failure,

- skin edema,

- high blood pressure,

- treatment with corticoids, estrogens or diuretics.

This diet consists of reducing sodium intake by eliminating the addition of salt (sodium chloride) in the kitchen, reducing or eliminating foods rich in salt, and sometimes using special foods without salt.

Permitted foods:

With about 1 g of salt/day, replace the basic food with its salt-free version:

Milk,

Deodorized bread,

Eggs,

Fresh meat and fish,

Dry vegetables,

Deodorized fats,

Deodorous pastry,

Lemon,

Spices,

Vinegar,

Fresh fruit,

Vegetables,

Low in salt,

Unsalted water

N.B.: After authorization of the doctor, replace the (sodium) salt by (potassium) salt.

Prohibited foods:

Table salt,

Common dairy products, especially cheese

Bread and rusks,

Salting

Pastries not deodorized, shellfish,

Deli meats,

Offal,

Smoked and dried fish,

Industrial preserves,

Chocolate and candy,

Effervescent drugs,

Sparkling water (Oulmes),

Celery,

Turnip,

Cabbage,

Salty condiments: mustard, pickles, olives

Diet : No residue or no fiber

The goal of this diet is to temporarily reduce or eliminate high-fiber foods that lead to the formation of a large amount of fecal matter (stool).

It is prescribed in cases where the mobility of the intestine is reduced:

-       Before and/or after digestive surgery,

-       In preparation for digestive examinations requiring a completely empty digestive tract (e.g. colonoscopy)...*etc.*

It is also intended for individuals with :

-       Digestive stomies,

-       Diarrhea

-       Various colitis

Permitted foods:

Grilled white meat,

Boiled fish,

Raw fats,

Yoghurts, white cheeses, plain or flavoured small Swiss cheeses but without fruit,

Rusk,

Fruit jelly,

Hard-boiled egg,

Homemade soup,

Prohibited foods:

Starchy foods,

Foods with too much sugar,

Bread,

Milk,

Delicatessen,

Raw fruit,

Dry and green vegetables,

Meat with fiber,

Fries and grease,

Sauces,

Soft drinks

Exclusion or avoidance diets in case of food allergies

Gluten-free diet

It is prescribed in cases of celiac disease, which is a disease of the intestinal mucosa that leads to malabsorption. In contact with gluten (allergen), a protein from wheat, barley, oats and rye, the intestine undergoes major alterations responsible for significant diarrhea and malnutrition.

Food deleted:

All wheat flour based foods (containing gluten protein)

Authorized cereals :

Corn,

Rice,

Oats,

Soybeans

Peptic Ulcer Diet

This diet is indicated in cases of ulcers, which are lesions in the gastric (stomach) and/or duodenal digestive mucosa caused by the *Helicobacter pylori* bacteria and by the regular consumption of non-steroidal anti-inflammatory drugs NSAIDs, such as aspirin or ibuprofen The diet is based on the <u>elimination of</u> foods that acidify the food bowl or increase digestive secretions:

Citrus fruits,

Alcohol,

Tobacco,

Spices,

Coffee

C. Other plan

<u>Diet : Hyper-protein/hypercaloric</u>

It is prescribed in case of pressure sores and malnutrition.

Permitted foods: (Example)

Breakfast

1 bowl of whole milk,

80g of cereals or 80g of flute + 2 teaspoons of butter + 2 tsp. jam

1 fruit or 1 glass of fruit juice

Lunch

200 g of white meat (red meat once a week) or 250 g of fish or 4 eggs, as many vegetables as you like (at least 100 g),

300 g of starch (8-10 tbsp.)

40 g of cheese + 1 to 2 slices of bread,

Snack: 1 semi-skimmed or whole milk dairy,

1 sweet product

1 to 2 slices of bread or equivalent.

Dinner

200 g of meat or equivalent,

Vegetables at will + 1 tablespoon of oil or equivalent,

300 g of starches (8-10 tablespoons),

1   fruit

<u>Miscellaneous remarks:</u>

-       Depending on the case or disease, medical diets may or may not be accompanied by regular physical activity.

-       Diets can have side effects:

1.   In the elderly, a strict deodorant diet can lead to dehydration and hyponatremia.

2.   In a syndrome of intestinal malabsorption, a diet without residue that is too severe, not very varied, monotonous, can lead to a reduction in intake and progressive undernutrition, and even certain elective deficiencies (vitamin C).

3.   Highly restrictive diets (e.g. low or very low calorie) are contraindicated in cases of progressive inflammatory or malignant disease, in cases of severe psychiatric disease or depression, and when there is a history of eating disorders.

References:

1.   Nutrition. Collège des Enseignants de Nutrition (Courses and MCQs for L2-L3 Medicine) (2014). Elsevier Masson SAS.

2.   Dietetics and Nutrition. Apfelbaum *et al* (2004). Abridged, 6th edition. Masson.

3.   Small Precises of Nutrition. Nicole Baudat (2008). Editions Lamarre.

4.   Rancé & Bidat (2006). Avoidance regimes: for whom and how? Avoidance regimes: for whom and how? Revue Française d'allergologie et d'immunologie clinique.

5.   IP2 Nutrition course at the Tiznit annex (also including nutritional habits in Morocco).

6.   Course: prescribing a diet from the French Virtual Medical University

Useful websites:

http://www.sante.gov.ma/Publications/GuidesManuels/Documents/Lutte%20contre%20les%20troubles%

5B1%5D.pdf

http://www.fao.org/food/nutrition/fr/

http://www.who.int/nutrition/fr/

http://www.wikipedia.com

https://vitadiet.net/

http://www.doctissimo.fr

yes
I want morebooks!

Buy your books fast and straightforward online - at one of world's fastest growing online book stores! Environmentally sound due to Print-on-Demand technologies.

Buy your books online at
www.morebooks.shop

Kaufen Sie Ihre Bücher schnell und unkompliziert online – auf einer der am schnellsten wachsenden Buchhandelsplattformen weltweit! Dank Print-On-Demand umwelt- und ressourcenschonend produziert.

Bücher schneller online kaufen
www.morebooks.shop

Printed by Books on Demand GmbH, Norderstedt / Germany